THE LITTLE BOOK OF
BUTTERFLIES

With color illustrations by Tugce Okay

ANDREI AND ALEXANDRA
SOURAKOV

PRINCETON UNIVERSITY PRESS
PRINCETON AND OXFORD

Published in 2024 by Princeton University Press
41 William Street, Princeton, New Jersey 08540
99 Banbury Road, Oxford OX2 6JX
press.princeton.edu

Library of Congress Control Number 2023943756
ISBN 978-0-691-25174-5
Ebook ISBN 978-0-691-25175-2

Typeset in Calluna and Futura PT

Printed and bound in China
1 3 5 7 9 10 8 6 4 2

British Library Cataloging-in-Publication Data is available

This book was conceived, designed, and produced by UniPress Books Limited

Publisher: Nigel Browning
Managing editor: Slav Todorov
Project development and management: Ruth Patrick
Design and art direction: Lindsey Johns
Copy editor: Caroline West
Proofreader: Robin Pridy
Color illustrations: Tugce Okay
Line illustrations: Ian Durneen

IMAGE CREDITS:
Alamy Stock Photo: 130 Sipa US; 134 Art Heritage. **Dreamstime.com:**
142 Susan Hodgson. **iStock:** 121 onebluelight. **Nature Picture Library:**
15l Nick Hawkins; 20 Konrad Wothe; 26 Kim Taylor; 29 Vladimir Medvedev;
40 Piotr Naskrecki; 48 Sven Zacek; 65 Thomas Marent; 66 Paul Aniszewski;
93 Hans Christoph Kappel; 110 Andy Sands. **Shutterstock:** 15r Josef
Stemeseder; 39, 151 Sari ONeal; 56 aspektreich Fotografie; 80l Matee
Nuserm; 80r Robert Ross; 83 KRIACHKO OLEKSII; 94 Cornel Constantin;
105 Danita Delimont; 124 Nigel Jarvis. **Other:** 71 sneak-e; 113 moniquayle.
Additional illustration references: 145 aacocucci.

Also available in this series:

THE LITTLE BOOK OF
SPIDERS

THE LITTLE BOOK OF
BEETLES

THE LITTLE BOOK OF
TREES

Coming soon:

THE LITTLE BOOK OF
FUNGI

THE LITTLE BOOK OF
WEATHER

THE LITTLE BOOK OF
DINOSAURS

THE LITTLE BOOK OF
WHALES

CONTENTS

INTRODUCTION

This book is a culmination of 27 years of collaboration between the authors. The first author's interest in butterflies began at the age of six and, over the last half-century, he has attempted to understand how butterflies and moths work, while never ceasing to be fascinated by their beauty. The second author accompanied her father from an early age on lepidopterological trips and spent countless hours with him at the museums where he worked or visited. In high school and college, she took these pursuits to the next level by introducing tools such as chemical ecology and materials science to the study of butterflies. Her career today is in engineering, but she continues to be fascinated by butterflies.

WHY YOU SHOULD READ THIS BOOK
(IF YOU'RE STILL ON THE FENCE)

After the writing of this book was completed, the *New York Times* ran an article entitled "She Can See the World in Colors He Can't Even Imagine," which spotlighted a study featured in the prestigious *Proceedings of the National Academy of Sciences*. It turns out that the females of the Zebra Longwing butterfly can see in the UV spectrum while the males cannot—all due to a single gene that has relocated in this species to a tiny sex chromosome absent in males. Curious, intriguing, bizarre . . . so much of the natural world reveals discoveries and phenomena that inspire these adjectives. The metamorphosis of butterflies from caterpillar to winged wonder has especially captured the imagination of mankind for centuries and given rise to many spiritual beliefs and allegories. The butterfly world provides a cornucopia of fascinating interactions, and the *New York Times* covers studies like these for the same reason that we wrote this book: these creatures are not only beautiful but engrossingly complex and diverse as well.

ABOUT THIS BOOK

We hope that this book will help the reader appreciate the duality of butterflies: as biological wonders with an intricate evolutionary history and adaptive behaviors, and as living works of art that have come to symbolize everything from the human soul to the ephemerality of life. As we show in the  first two chapters, butterflies precede us by 100 million years and have persisted through the rise and fall of many species, adapting to a variety of habitats, even some of the most extreme ones. This span of time has allowed for many divergences in the evolutionary tree, resulting in the seven butterfly families that we profile in Chapters 3 to 5. Chapters 6 to 9 delve into butterfly physiology and their ecological interactions. Chapters 10 and 11 cover human interactions with butterflies, from conservation issues to folklore. The book concludes by highlighting some of the remarkable behaviors exhibited by butterflies.

We hope that through this book, the reader will find, in the words of Oscar Wilde, "beautiful meanings in beautiful things." For those who are scientifically inclined, there is detailed discussion of biology, evolution, and physiology, while the panoply of illustrations and photographs celebrates the butterfly's beauty and diversity. We also hope to draw attention to the dangers facing butterflies. Their decline is tolling a warning bell to which we may want to be sensitive, for their sakes and our own.

Andrei and Alexandra Sourakov

WHAT IS A SPECIES?

There are around 19,000 described butterfly species, most of them found in the tropics. A "species" is frequently defined as a group of organisms that can mate to produce fertile offspring. In practice, applying this definition to two populations of butterflies can be tricky. How can one test the hypothesis that two populations belong to two different species, when there are normally only a few dead museum specimens to reference? Luckily, scientists can draw on secondary evidence when delineating a species, such as morphological differences in internal and external organs and comparisons of DNA sequences.

EVOLUTION OF TAXONOMIC METHODS

Once scientists realized that wing patterns can be variable within a species, they developed a method of dissecting and drawing male genitalia. These organs frequently differ across species but prove to be much more consistent than wing patterns within a species. Today, these characteristics are frequently integrated with knowledge of a species' natural history, distribution, and DNA to inform classification.

↓ The dissecting microscope is one of the main tools used by butterfly taxonomists to study specimens.

↓ Male genitalia can help distinguish between species when wing pattern characteristics are confusing.

→ Closely related species, such as these Neotropical satyrines—here (A) *Pierella helvina*, the red-washed satyr, (B) *P. hyceta*, (C) *P. lamia*, (D) *P. lena*, and (E) *P. nereis*—are grouped into genera. There are many different species with wing patterns like those of *P. lamia* but the genitalic structures of each are unique.

DESCRIBING DIVERSITY

Different species concepts were prevalent at different times in history, but the work of taxonomists is cumulative, meaning earlier descriptions and species' names are as important today as the newer ones.

The main taxonomic categories within the order Lepidoptera are family, genus, and species. There is also a frequent need for intermediate-level categories above the genus but below the family that group species by shared characteristics. Names of different categories have specific endings: "-ini" for tribes (e.g., Ithomiini, the clearwing butterflies); "-inae" for subfamilies (e.g., Danainae, to which Ithomiini plus Danaini, the milkweed butterflies, belong); "-idae" for families (e.g., Nymphalidae, or brush-footed butterflies); and "-oidea" for superfamilies (e.g., Papilionoidea).

FROM ARISTOTLE TO MAYR

Detailed records of plants and animals go back to the time of Aristotle (384–322 BC), who assembled large collections and made systematic observations of living things. In the middle of the 18th century, the Swedish naturalist Carolus Linnaeus (1707–1778) introduced binomial nomenclature, giving each species a Latin genus-species name.

~ The first compendium ~

Linnaeus's *Systema Naturae* was one of the first attempts at a comprehensive overview of all plants and animals. The tenth edition, in which he introduces zoological nomenclature, recognized just one genus for all butterflies (*Papilio*), into which he placed nearly 200 species (only 18 of which are still categorized in the Papilionidae family). Johan Christian Fabricius (1745–1808) continued to chisel away at the monumental task of describing insect diversity and established the modern insect classification system.

↓ Fabricius named 9,776 insect species. He also subdivided described species into orders and genera, laying the foundation for modern classification.

It took another hundred years to develop the "biological species" concept, which is prominent in today's textbooks and which was first proposed in *Systematics and the Origin of Species* by Ernst Mayr (1904–2005). Using this concept, scientists approach species as groups of constantly evolving, interbreeding populations, rather than individuals defined by their unique appearance or behavior. It is also widely recognized now that different species can form "hybrid offspring" (products of inter-species mating, usually with lower fitness) that sometimes facilitate gene flow across species boundaries.

A SPECIES IS BUT A HYPOTHESIS

While evidence from morphology and DNA can tell us a lot about a species, the population structure, distribution, and behavior are still only known for a tiny portion of butterflies. Since we are constantly learning new information, existing species designations can be viewed as a hypothesis that is tested every time conflicting evidence is discovered. Scientists are constantly describing new species and synonymizing old ones, as disparate pieces of evidence emerge and are reconciled. Certainly, with some 19,000 described butterfly species and numerous subspecies, there is still a lot of work ahead!

IMPORTANCE OF SUBSPECIES

Two allopatric (geographically non-overlapping) species may produce hybrids in their contact zone (if one exists), but gene flow is severely limited, causing these populations to diverge more and more. However, if such butterfly populations remain completely reproductively compatible with each other, subspecies are frequently named, making their scientific names "trinomial."

THE CASE OF THE WHITE ADMIRAL

The White Admiral butterfly is distributed throughout most of North America (except in the southeast) and is known by its scientific name, *Limenitis arthemis arthemis*. Its relative, the Red-spotted Purple, *L. arthemis astyanax*, is found from the Great Lakes in Michigan to Florida and has evolved a color pattern that mimics the toxic Pipevine

---

SUBSPECIES AND CONSERVATION

While some scientists find the subspecies concept too vague and argue against it, others stress that by officially describing a subspecies, not only is the phenotypic but also the genetic diversity contained within a species preserved. Formally recognized populations tend to attract better protection from conservation biologists and the public. For example, two butterflies that are listed under the Endangered Species Act in the United States, the Schaus' Swallowtail, *Papilio aristodemus ponceana*, and the Miami Blue, *Cyclargus thomasi bethunebakeri* (pictured), are unique-looking subspecies that have been protected by this status. Having said that, there are as many opinions on species and subspecies concepts as there are taxonomists.

Swallowtail, *Battus philenor*. The difference in their appearance is due to mimicry alone and nothing prevents gene flow between these two reproductively compatible populations. As a result, they hybridize in geographic regions where their ranges overlap and produce intermediate-looking forms.

EN ROUTE TO "SPECIES"

A subspecies can be a stepping stone to forming a new species. When two populations become isolated by a geographic barrier, they initially maintain the ability to interbreed with one another, but the opportunity to do so is limited due to lack of contact. As time goes on, such populations tend to accumulate significant enough differences through mutation and natural selection. Even if they were to overcome the geographic barrier, they would no longer be able to interbreed—they are now separate species.

↑ Both the White Admiral butterfly and the Red-spotted Purple are subspecies of the same species, *Limenitis athemis*, which split into two distinct populations around 230,000 years ago but which are now hybridizing in the northeastern United States.

POLYMORPHISM

When two or more forms coexist in equilibrium within a population, it is known as "polymorphism." Unlike the subspecies concept, polymorphism does not have a significant geographic component. The most common form is sexual dimorphism, in which males look different from females, but there are other types as well. Among females of the Tiger Swallowtail, *Papilio glaucus*, in the eastern United States, the dark form mimics Pipevine Swallowtails and may be avoided by birds, while the more widespread tiger-striped form may help escape notice by predators. Neither form has gained the upper hand in survival, and thus both persist in the same population.

SEASONAL POLYPHENISM

Many butterflies have wet-season and dry-season forms, the latter usually being more cryptic (see Chapter 9, page 117). But one of the most spectacular examples of seasonal polyphenism is the mimetic European Map Butterfly, *Araschnia levana*. This butterfly feeds as a caterpillar on nettles and produces a spring generation that is red with black spots, resembling distasteful checkerspot butterflies, such as *Melitea* and *Euphydryas*, and a summer generation that is black with white stripes, which resembles a tiny white admiral (*Limenitis* species).

↓ Seasonal polyphenism is shown by the European Map Butterfly, which has two entirely different forms that appear in consecutive generations. Environmental factors, such as day length and temperature, cause the same genes to produce different phenotypes.

→ Two forms of the female Tiger Swallowtail are found in the southeastern United States, occurring within the same populations.

WHEN DID BUTTERFLIES EVOLVE?

The seven families discussed in detail in Chapters 3 to 5 form a monophyletic group, meaning they are all descended from a common ancestor. Although the oldest known butterfly fossil (*Protocoeliades kristenseni*, found in marine deposits in Denmark in 2016 and described as a new, now-extinct genus and species of skipper butterflies) dates back 55 million years, it does not look drastically different from modern butterflies. Scientists suspect, however, that the first butterflies are actually twice as old as that.

EVOLVING FROM MOTHS

Insects are grouped into orders according to various morphological features and are especially differentiated by mouthparts. Butterflies are most closely related to moths, and together they form the order Lepidoptera (meaning "scaled wings"). Their closest relatives are the caddisflies (Trichoptera), which resemble moths as adults but have aquatic larvae. Lepidoptera split from Trichoptera close to 300 million years ago (Mya). These early Lepidoptera initially fed on pollen using mandibles, which some moths still do, and evolved terrestrial,

---

EVOLUTION: HOW DO WE KNOW?

We know from the fossil record that the mass extinctions of some life forms are followed by dramatic change and radiation, like the burgeoning of birds after the dinosaurs. Yet many life forms, such as crocodiles, have not changed drastically for close to 200 million years. What we know about the evolution of Lepidoptera comes from a combination of the few available fossil records, the morphology of existing species, and from their DNA. Butterfly fossils are up to 60 million years old, but some moth fossils formed over 175 million years ago. Fossils help "calibrate" DNA-based evolutionary hypotheses.

---

rather than aquatic, larval forms that fed on plants. It took another 100 million years to develop the hallmarks of today's Lepidoptera: the sucking, tube-like mouthparts, known as the proboscis (see Chapter 7, page 85 and Chapter 8, page 96).

EVOLVING MATING SYSTEMS AND PUPAE

Another big leap in Lepidoptera evolution was a reproductive system in which the openings for mating and for egg laying are separate. All butterflies, and most moths, share this characteristic and are grouped under Ditrysia. At some point, over 100 Mya, some moths (and from them, butterflies) also evolved fused abdominal segments in the pupal stage, forming the group Obtectomera. Contrary to popular belief, butterflies are not the final word in Lepidoptera evolution: some 88,000 species of so-called "eared moths," comprising many families, have evolved following the emergence of butterflies.

↓ *Prodryas persephone* is a now-extinct fossil butterfly, which has been found in Colorado between thin layers of sedimentary rock dating back to 34 Mya.

THE FIRST BUTTERFLIES

For the most part, ancient insects are best preserved in amber deposits, and while some can be found in rock formations, there are few well-preserved rock fossils. Based on the butterfly fossil evidence we do have and comparative morphological and DNA studies, many scientists believe that Papilionidae (swallowtails) is the most ancient lineage of butterflies, followed by Hesperiidae (skippers) and Hedylidae (American moth-butterflies). All three are discussed in detail in Chapter 3.

TRAPPED IN TREE SAP AND BURIED IN ROCK

Rock fossils of extinct butterflies have been found in Florissant, Colorado, a site estimated to date back 34 million years. These species, as well as the ones discovered in amber, shed light on the differences between extant and extinct species, but they are often not as different as we might think. A 15,000-year-old fossil from Colombian copal

(amber's younger form) looks identical to a modern butterfly. Specimens found in Dominican amber dating back 15–25 million years represent an extinct species of Riodinidae, *Voltinia dramba*. Scientists think that these butterflies may have fed on bromeliads and were trapped in tree sap while laying eggs. A *Voltinia* species morphologically similar enough to be placed in the same genus can still be found in Mexico today.

PLANT RADIATION AND BUTTERFLY EVOLUTION

Around the same time that the first butterflies evolved, many modern plant families radiated and, in the process, evolved a plethora of toxic chemicals to protect themselves from herbivores. Perhaps, the ability of some moth caterpillars to take advantage of toxic plant compounds holds the key to understanding the early evolutionary steps of butterflies. By becoming chemically defended, butterflies may have been able to fill an ecological niche: flying during the day and pollinating the rapidly diversifying, day-blooming flowers.

← The Six-spot Burnet, *Zygaena filipendulae* (left), a European moth in the family Zygaenidae, and the Large Skipper butterfly, *Ochlodes sylvanus*, caught in the same frame.

WHERE CAN BUTTERFLIES BE FOUND?

The answer is: pretty much everywhere if you know how to look. Butterflies are highly adaptable and their life cycle allows them to adjust to a wide variety of climatic conditions. As a result, we can find butterflies occupying habitats that to us appear extreme (see page 26).

CITIZEN SCIENCE AND BUTTERFLIES

When one compares the 60 butterfly species found in the United Kingdom, or the 750 species living in the enormous territory of the United States, with the nearly 2,000 species described in one square mile of the Amazonian rainforest in Brazil, one may find it ironic that until recently most of the information about butterflies came from people who lived in Europe, Japan, or North America. Fortunately, this is rapidly changing, as countries such as Brazil and many others develop their own butterfly research programs. Additionally, citizen science initiatives help promote interest in butterflies and organize the collection and exchange of information about these insects.

↓ Today, observations one makes can be uploaded and accessed via the iNaturalist website, which also generates maps of animal and plant distributions, providing valuable information for research scientists.

→ The Small Apollo, *Parnassius phoebus*, is found in Alaska and above the tree line in the European Alps. In the Himalayas and Central Asian mountains, other species of *Parnassius* can fly even higher, above 13,000 ft (4,000 m). Stonecrops, the succulent host plants of these butterflies, are well-adapted to growing among rocks at high elevations.

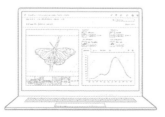

TIED TO THEIR HOST PLANTS

The lives of butterflies are intimately connected to their host plants. After over 100 million years of coevolution, butterflies can thrive on a patch of sparse grass in the desert, feed on the tough leaves of towering oak trees, and derive nutrients from soggy mosses in a dark rainforest.

Many butterfly species manage to use the toxic chemicals, originally evolved by host plants as protection against herbivores, for their own benefits, such as nutrition, defense, and reproduction. While using secondary plant compounds has many benefits, it is not without some cost: some butterfly species become so specialized that they can no longer survive without a specific plant. Frequently, female butterflies will not lay eggs and caterpillars will not feed on anything but their chosen host species.

ADDICTED TO ACIDS

Many groups of plants have developed specialized compounds to defend themselves against microbes and insects, but some butterflies have successfully made these defenses their own. Tropical swallowtails frequently feed on pipevines (*Aristolochia*), detoxifying their

---

A TASTE FOR PINE

In the mountains of western North America, two species of pine whites (*Neophasia*) can experience population explosions when their coniferous host plants thrive. But try chewing on a pine needle (as the *Neophasia* caterpillar is doing here) and you will soon realize that to master this diet, these butterflies had to develop some pretty serious detoxification abilities. Pine oils contain over 20 defensive compounds, such as a-Terpineol and limonene.

---

aristolochic acids (see Chapter 3, pages 38–39) and using them to deter their own predators. Close to 30 species in the genus *Telipna* (Lycaenidae) can feed as larvae on species of mosses known to contain oxylipins and terpenoids in the tropical forests of West Africa. These detoxification abilities arise through changes in DNA that lead to the modification of chemical pathways (see Chapter 6, page 81).

CHEWING ON GLASS

Grasses are tough and poor in nutrients, which are primarily concentrated in seedheads, but they are hardy and among the first to colonize poor soils. Over 2,000 species of Satyr butterflies (tribe Satyrini) have taken advantage of the wide availability of grasses, and their cryptic caterpillars can be found hidden in plain sight, aligned along narrow leaf blades of grasses and sedges throughout the world. Many satyrs also feed on bamboos, which can grow within virgin rainforests. Rich in silica (SiO_2), an inorganic constituent of glass and sand, these grasses and bamboos are tough and quickly wear down the mandibles of hungry insects. As caterpillars grow, they are able to renew their mandibles once they molt, but the tougher, nutrient-poor diet of satyrines means that they have to molt more often and develop more slowly.

EXTREME HABITATS

Butterflies are very good at locating their host plants and nectar sources, even in the most inhospitable environments, like big cities. Yes, even within bustling, crowded New York City, one can find butterflies. For animals that are accustomed to surviving in the most hostile conditions, including icy mountain peaks above 15,000 ft (4,500 m) in elevation, coastal marshes inundated by saltwater tides, and deserts scorched by the sun, skyscrapers and asphalt are just another challenge to adapt to.

FROM DESERT TO TUNDRA

The short but explosive blooming season that occurs in the desert after the rains is a great time for butterfly-spotting. For instance, the Desert Sand-skipper, *Croitana aestiva*, which is restricted to the arid region of Australia's Northern Territory, times its reproduction to coincide with the downpours.

LIVING IN THE LION'S DEN

Few butterflies are carnivorous, but one particular group within the Lycaenidae family has established this niche for itself. Several species of polyommatine blues can be found developing inside ant nests, either tricking the ants into feeding them or directly consuming the ant's progeny. Living with such dangerous hosts requires chemical finesse; the caterpillars need to send just the right signals to deceive the ants into being accepted as one of their own.

Hiking above the tree line, you might see butterflies, such as the *Parnassius* species, flying close to the snow, often just for a brief moment when the sun comes out. Some species have adapted to endure their entire life cycle in the tundra, while others come up just for the season. For example, the sedentary Melissa Arctic butterfly, *Oeneis melissa*, requires two years to complete its reproductive cycle in the harsh Arctic north, while the migratory Painted Lady, *Vanessa cardui*, repopulates the Arctic region from its winter breeding grounds in Africa.

SURVIVING THE FLOODS

One environment that insects have not mastered is the open ocean. However, the Eastern Pigmy Blue, *Brephidium pseudofea*, one of the smallest butterflies in the world, can survive immersed in salt water as eggs and pupae for up to 20 days, thanks to specialized air pockets that trap oxygen. As caterpillars, they feed on the perennial glasswort in the coastal marshes of the eastern United States.

← The Pioneer White butterfly, *Belenois aurota*, thrives in desert environments, even in the Kalahari desert, in southern Africa. Its migratory behavior allows it to follow sporadic rains and occupy vast territories.

GEOGRAPHY AND EVOLUTION

Butterflies branched off from moths 100 Mya, on a land mass that today is part of the New World. The proximity of land masses in that age allowed butterflies to radiate quickly, diversifying first in the tropical regions of these lands and then adapting to more temperate climates. As continents moved farther apart from each other, the butterfly fauna of each Zoogeographic Region began to chart its own evolutionary course.

ZOOGEOGRAPHIC REGIONS

The 19th-century naturalist, Alfred Russel Wallace (1823–1913), contributed greatly to the understanding of speciation and the past and present geographic distribution of animal species. In addition to coauthoring the theory of natural selection with Charles Darwin (1809–1882) in 1858, Wallace outlined the Zoogeographic Regions that we use today when discussing animal diversity: the Palearctic, Afrotropic, Indomalayan, Australasian, Nearctic, and Neotropical. These regions are delineated based on distinctive fauna united by a common evolutionary history. Wallace's original demarcation

VICARIANCE VERSUS DISPERSAL

Opinions on the mechanism of early butterfly evolution used to be polarized, with one camp favoring vicariance (evolution driven by continental drift) and the other favoring dispersal, believing that bodies of water were not a sufficient obstacle for butterflies. It turns out that both sides are right, as the evidence points to some butterfly groups having a mark of vicariance on their evolutionary history, while other lineages can only be explained by periodic dispersal. Continents were much closer together when early lineages evolved about 100 Mya, so dispersal was probably a much more frequent event.

↑ Found in both eastern Eurasia and northwestern North America, the distribution of *Parnassius eversmanni* is the result of changing sea levels and climate during the Pleistocene.

was drawn from direct observations and records of animals, which have since been bolstered by genetic, morphological, and geographical analysis.

CONTINENTAL DRIFT

Zoogeographic Regions roughly align with the major land masses, and it was not lost upon Wallace or his contemporaries that the continents fit almost like jigsaw pieces, not just in their shape but geologically as well. Alfred Wegener (1880–1930) first used the term "continental drift" when he presented his theory—that the land masses had once all been conjoined and had "drifted" apart over time—in 1912, but widespread acceptance was stymied by the lack of a convincing explanation for what was causing this drift. The science of plate tectonics would go on to provide such an explanation in the second half of the 20th century.

EVOLVING IN ISOLATION

Many islands and continents have endemic butterfly species, meaning that they occur there and nowhere else, but many species span the globe. In fact, for butterflies that have powerful flight, it has been possible to document long-distance, overwater dispersal in real-time. Well-documented migrations of the Monarch, *Danaus plexippus*, and the Painted Lady, *Vanessa cardui*, show that some species can routinely overcome great distances and barriers (see Chapter 8, page 103 and Chapter 12, page 142). However, for many weak-flying butterfly species, even a relatively narrow barrier like the Amazon River is sufficient to isolate them.

ISLAND SPECIATION

Since Darwin's observations of Galapagos finches, islands have been viewed as ideal places for understanding speciation, including that of butterflies. Excellent examples of this phenomenon include the slow-flying satyrine butterflies of the genera *Calisto* and *Mycalesis* that inhabit the Caribbean and the Solomon Islands and the fast-flying birdwings (*Ornithoptera*) that have evolved several spectacular species and subspecies across the Pacific Islands.

↓ (A) Cuba, (B) Jamaica, (C) Hispaniola, (D) Puerto Rico. These islands were more fragmented 30 Mya, when the genus *Calisto* originated, so it was easy for populations to become isolated.

→ (A) *Calisto herophile*, Cuba, (B) *C. zangis*, Jamaica, (C) *C. hysius*, Hispaniola, (D) *C. nubila*, Puerto Rico. These 4 butterflies represent the 40 or so *Calisto* species of the Caribbean. They may look superficially similar but have different anatomies, and their DNA suggests ancient radiation events.

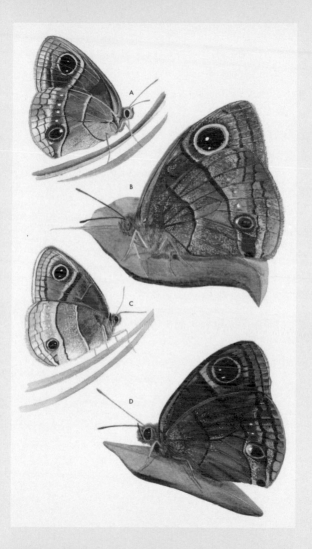

ELEVATION AND DIVERSITY

If one maps species richness per square mile on the South American continent, it becomes clear that there are many more butterfly species in the Andes than in the Amazon. Mountains in the Nearctic region, such as the Rockies, Pamirs, Tian Shan, Alps, and Pyrenees, also have many more butterfly species compared to the surrounding flat lands. For example, a dozen species of *Euphilotes* blues are recognized in mountainous regions from California to Colorado and close to 100 species of ringlets in the genus *Erebia* are found throughout mountains of the Holarctic realm (all habitats of the Northern Hemisphere that lie north of the Neotropics, Afrotropics, and Southeast Asia).

THE FOUNDER EFFECT

As the glaciers were receding after the last North American ice age, wood-nymphs (*Cercyonis* species) developed into numerous unique-looking populations with different patterns and colors. Over 100 subspecies have been named within this small genus. The Great Basin Wood-nymph (*C. sthenele*), Mead's Wood-nymph (*C. meadii*), and the Small Wood-nymph (*C. oetus*) are restricted to mountain habitats in the western United States, while the Common Wood-nymph (*C. pegala*) spans the entire country.

In the eastern United States, the Common Wood-nymph provides a perfect illustration of clinal variation, a phenomenon in which disjunct populations vary ever so slightly but end up looking like different species at the extremes of the species' distribution, due to the compounding differences that accumulate across its range. The "founder effect"—when an isolated population carries the features of the single female butterfly that gave rise to it—may be responsible for this diversity, in combination with variable selective pressures that further shape populations.

EVOLVING IN THE MOUNTAINS

Mountain chains create very distinctive habitats, such as coastal dry forests and cloud forests, by influencing climate. Studies of butterfly faunas in countries like Costa Rica, Colombia, Ecuador, and Peru suggest that they are extremely species-rich, despite their relatively small size, because their mountains serve as hubs for numerous ecosystems. They create isolated habitats for butterfly diversification, as the lowland dwellers rarely mix with denizens of higher elevations.

↑ The number of chromosomes can change through fragmentation or fusion. During speciation within *Erebia*, such changes may have contributed to the evolution of nearly 100 species.

SPREADING FROM MOUNTAINS TO PLAINS

It is true that some of the richest butterfly fauna has been documented at lower elevations in the South American tropics, such as in the Tambopata valley of Peru or Rondônia in Brazil. Indeed, the rainforests offer numerous niches for animals to co-occur within the same habitats, as they support prodigious plant diversity and many microhabitats created by the forest canopy. Yet, the divergence of many of these species has been tied to the Andes, the mountain chain that spans almost the entire South American continent. One can view the mountains as the speciation hubs and the rainforests as the sponges ready to absorb species that originate in the mountains.

GOING NORTH BY GOING UP

In addition to driving evolution through isolation, mountains also create more ecological niches for species to inhabit by harboring habitats with distinct climates and flora. This was first described comprehensively by Alexander von Humboldt (1769–1859), an explorer who climbed the mountain of Chimborazo in the Andes, at the dawn of the 19th century, reaching almost 20,000 ft (6,096 m) of altitude. He wrote that, as he ascended, it was as if he were traveling northward from the equatorial jungles all the way to the Arctic tundra.

SWALLOWTAILS (PAPILIONIDAE)

The swallowtail family has over 500 species of the most spectacular diversity—some enormous and some small, each adapting to a different lifestyle. However, all swallowtails have the same arrangement of wing veins, a feature frequently used in the classification of Lepidoptera. Their caterpillars also share common features, like stout bodies and defensive osmeteria, glands that are displayed when the caterpillar is disturbed, producing a repellent smell (see page 37).

BUTTERFLY FOSSILS

If a "proto-butterfly" fossil is ever found, it would likely be placed in the swallowtail family. The latest evolutionary tree suggests that butterflies branched off the much larger and significantly older moth lineages about 100 Mya.

Some of the older fossils discovered to date are specimens of the extinct genus *Praepapilio* that date back to the mid-Eocene (approximately 45 million years old). They are more closely related to the subfamily Baroniinae (see page 36) than to the rest of the swallowtail family, linking these fossils to the oldest surviving swallowtail lineage.

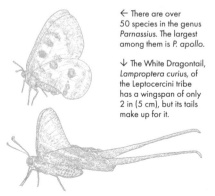

← There are over 50 species in the genus *Parnassius*. The largest among them is *P. apollo*.

↓ The White Dragontail, *Lamproptera curius*, of the Leptocercini tribe has a wingspan of only 2 in (5 cm), but its tails make up for it.

→ *Baronia brevicornis*, commonly known as the Short-horned Baronia, belongs to the most ancient lineage among surviving swallowtail species. It has several color morphs, with orange females being the rarest. As a caterpillar, it feeds on the boat-spine acacia (*Vachellia campechiana*) found in the dry mountains of western Mexico.

SWALLOWTAIL DIVERSIFICATION

In evolutionary terms, swallowtails are the eldest of the seven butterfly families, which has given them a chance to develop a great diversity in appearance and life history, as well as an aptitude for detoxifying toxic plants.

ANCESTRAL LINEAGES

The oldest ancestral lineage of Papilionidae has only one surviving species: *Baronia brevicornis*. It lives in the dry forests of the Sierra Madre in Mexico, where its slender and cryptic caterpillars construct shelters by pulling together the leaves of their host plant, the boat-spine acacia. When the caterpillars are ready to pupate, they drop to the ground and bury themselves in the soil, a behavior more typical of moths and skipper butterflies.

DIVERSITY OF APOLLO BUTTERFLIES

The next-to-evolve lineages produced *Luehdorfia* (comprising four Asian species) and *Parnassius*, a genus known as the Apollo butterflies. Many of these species feed on stonecrop (*Sedum*) host plants that thrive on mountain outcrops at high elevations, providing food for *Parnassius* larvae that pupate under rocks and overwinter as eggs. Some species, such as *P. mnemozyne*, fly in meadows in the lowlands and feed on *Corydalis*, herbaceous alkaloid-containing perennials, throughout a vast region of the temperate Palearctic.

KITES AND DRAGONTAILS

In addition to the three subfamilies, Baroniinae, Papilioninae, and the now-extinct Praepapilioninae, swallowtails are grouped into 31 tribes. One of them, Leptocercini, contains long-tailed kite swallowtails (for example, *Graphium* and *Eurytides*). The smallest swallowtails in this clade, dragontails of the genus *Lamproptera*, measure just over an inch in forewing span. Leptocercini are largely tropical, but some species, such as the Zebra Swallowtail, *Protographium marcellus*, reach as far north as the Canadian border, ditching pipevines for paw-paw host plants.

TREE-FEEDING CATERPILLARS

One member of the Papilionini tribe, the Homerus Swallowtail, *Papilio homerus*, is the largest butterfly in the Americas and is endemic to Jamaica. Its enormous caterpillars are associated with a tree in the genus *Hernandia*, which is also unique to Jamaica. The harvest of this tree for charcoal and other land developments threatens the existence of *P. homerus*, but several conservation campaigns have helped prevent its extinction.

Within the tribe Papilionini, there are several species, collectively referred to as tiger swallowtails, which are so similar to each other that they can hybridize, like the Canadian Tiger Swallowtail, *Papilio canadensis*, and the Eastern Tiger Swallowtail, *P. glaucus*. While they use many species of hardwood trees, such as cherries and magnolias, as host plants, they can be quite picky eaters. Studies have shown that each population within a species develops strong preferences for certain host plants, and their survival can be impacted by the absence of their favorites.

The caterpillar of the Asian Swallowtail, *Papilio xuthus*, exposes its osmeterium, releasing a repellent odor.

SWALLOWTAILS ON PIPEVINES

Overcoming the chemical defenses of pipevines (*Aristolochia*) and utilizing them for their own protection helped swallowtail genera in the tribe Troidini to diversify. As the pipevine plant spread across the globe, its butterfly subscribers followed. The chemical protection that the caterpillars gained from feeding on these toxic plants, with their aristolochic acids, enabled them to ward off predators and survive better than more palatable caterpillars.

MIMICS OF THE PIPEVINE SWALLOWTAIL

The chemical defenses of troidines are so potent that many other butterflies have evolved to resemble these pipevine-feeding swallowtails. For instance, the Pipevine Swallowtail, *Battus philenor*, in North America serves as a model for a host of edible butterflies, such as females of the Diana Fritillary (*Speyeria diana*) and the Red-spotted Purple (*Limenitis arthemis astyanax*), both of which are in the nymphalid family. Several members of their own family also mimic *B. philenor*, gaining some of the benefits without the work of sequestering aristolochic acids, including the Spicebush Swallowtail (*Papilio troilus*), females of the Black Swallowtail (*P. polyxenes*), as well as the female mimetic form of the Eastern Tiger Swallowtail (*P. glaucus*).

BIRDWINGS AND CATTLEHEARTS

Troidines exhibit great diversity in size, shape, and geographic distribution. Enormous and eye-catchingly shiny birdwings can be found in Austronesia, with a particular hotspot in New Guinea. The smaller cattlehearts (*Parides*) also sport an array of iridescent colors but are found on the other side of the world, with nearly 40 species throughout the South American tropics. No matter how different the pipevine-feeding genera look from one another in adulthood, their caterpillars and pupae remain remarkably similar, a testament to a design that has stood the test of time.

PLANT DEFENSE PARADOX

As Pipevine Swallowtail caterpillars munch on their host plant, the plant retaliates by increasing its toxicity. In fact, the toxins it produces can increase five-fold. One might think that this would be detrimental to the caterpillars, but they do not seem to mind and continue feeding and developing well, in spite of their spicier meal. On the other hand, the caterpillars' predators feel the effects— when they feed on caterpillars that have ingested larger amounts of aristolochic acids, they either spit them out or are poisoned.

↓ Young caterpillars of the Pipevine Swallowtail feed together in order to overcome plant defenses and simultaneously signal their toxicity to predators, while larger caterpillars feed alone.

AMERICAN MOTH-BUTTERFLIES (HEDYLIDAE)

Hedylidae represent only about 0.1 percent of butterflies, but they deserve a mention in this book because they are very important for understanding the evolutionary connection between butterflies and moths. Despite their overall moth-like appearance, they share some 20 morphological characteristics with butterflies.

IDENTIFICATION THROUGH TRIANGULATION

Currently, all 35 species of Hedylidae are placed into a single genus, *Macrosoma*. The author of the first life history of a species in this genus (*M. heliconiaria* from Mexico) experienced one surprise after another as he reared the caterpillars he had found.

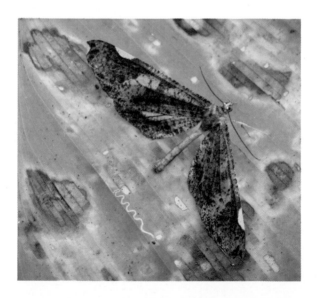

Based on the morphology of the slender, hairless caterpillars with prominent head horns, the author thought he was rearing a satyrine butterfly, a member of the family Nymphalidae. When the caterpillars pupated, however, he became certain that the species was actually part of another butterfly family, as the chrysalides were typical of Pieridae. So, when the hedylid "moths" (as they were thought to be at the time) popped out of these chrysalides, he was flabbergasted. Imagine his surprise if he were to learn that today this species is classified as a butterfly.

CROSS-CHECKING WITH DNA

Since then, these nocturnal butterflies have been studied in great detail. DNA sequencing and analysis has confirmed their position as one of the most ancestral of butterfly lineages, together with swallowtails and skippers. Another exciting discovery was their ultrasonic "hearing" abilities—a result of special organs located on the wings that evolved to detect ultrasound in order to avoid bat predators.

← *Macrosoma hyacinthina* is in the only genus representing the family Hedylidae, in which the immatures resemble butterflies, but the adults look and behave like moths.

SKIPPERS (HESPERIIDAE)

According to the latest DNA-based evolutionary hypothesis, around 90 Mya, skippers and hedylids parted ways with swallowtails, with the first distinct skippers appearing 20 million years later. There are about 3,500 described skipper species today, grouped into 11 subfamilies, with new species being added every year.

AS FAST AS A FLASH

For ecologists and evolutionary biologists, skippers are a treasure trove, as they demonstrate tremendous versatility and a variety of unique adaptations. Their fast reflexes and quick, jumping flight make it difficult for birds to catch them. If one takes a photo of a perching skipper with the flash on, it frequently reacts to the pre-flash in the camera so quickly that the butterfly will be captured in flight every time.

AFRICAN GIANT SKIPPER

There are exceptions to the speedy escape strategy, even among skippers (see page 45). Caterpillars of the African Giant Skipper, *Pyrrhochalcia iphis*, for example, feed openly in groups—a sure sign that they have other means of protection. Indeed, they feed on toxic plant families, such as Anacardiaceae (to which cashews and poison ivy belong), which can endow them with chemical defenses.

As adults, these skippers are brightly colored, reach 4 in (10 cm) in wingspan, and fly relatively slowly. This species is part of the subfamily Coeliadinae, the most ancestral among the skippers. Thus, it is possible that the unique features of this butterfly reflect the earlier evolutionary traits of skippers, with younger skipper species evolving a more compact size and faster flight to compensate for their lack of chemical defense.

ESCAPE MIMICRY

While most skippers are dark-colored and understated, some have iridescent colors on their upper wing surfaces, which they display readily when perching. In addition to functioning as signals to members of their own species, both male and female, the iridescent markings are thought to remind birds that their owners are too fast to be worth the chase.

Generally, toxic mimetic species fly slowly and signal to birds with their bright, conspicuous coloration: "Avoid eating me if you don't want an unpleasant taste in your mouth." In contrast, skippers engage in escape mimicry, conveying the message: "The last time you chased me, you wasted lots of time and energy and went hungry— it's not worth it." Mimicry among adult skippers results in their caterpillars differing from each other more than their adult forms do (see pages 44–45).

↓ A long proboscis allows skippers to reach nectar in deep flowers, but occasionally they will use it to feed on bird droppings, excreting some water to moisten their meal.

DIVERSITY OF SKIPPER CATERPILLARS

Skipper caterpillars frequently have eyespots on brightly colored heads or on the last abdominal segments. These spots and their body stripes can help scientists distinguish species that otherwise would be misleadingly similar. In fact, 10 different skipper species were once differentiated based on DNA and caterpillar morphology in Costa Rica, hiding under the same winged disguise.

UNIQUE DEFENSES

Skipper caterpillars are edible to a wide variety of predators and have hence developed a set of unique defenses. For example, they frequently build shelters, either by pulling together leaf margins or by cutting out parts of the leaf and using them as building material, making extensive use of silk thread. Yucca skippers of the genus *Megathymus* burrow into the roots of their host plants and create silk tunnels underground. The caterpillars of the Brazilian Skipper, *Calpodes ethlius*, which can be found from the United States to Argentina feeding on canna lilies and alligator-flag plants, have translucent skin, so that one can observe their internal organs at work—if you can spot the caterpillars!

↓ Skipper caterpillars, such as this Brazilian Skipper on an alligator flag plant (*Thalia*), make shelters by cutting and folding the leaves, which they secure with silk. To avoid predators, the caterpillars come out to feed at night.

→ (A) Brazilian Skipper, (B) African Giant Skipper, (C) Long-tailed Skipper. Among these species, only the caterpillars of the African Giant Skipper feed in the open during the day. They are colored aposematically and sometimes feed in groups, indicating that they are probably toxic to predators.

WHITES AND SULPHURS (PIERIDAE)

About 75 Mya, the lineages of white and sulphur butterflies began diversifying. Together with the highly mimetic Neotropical tribe of Dismorphini butterflies and a small genus *Pseudopontia* found in West Africa, they form the family Pieridae. Today, this family consists of over 1,000 species grouped into over 70 genera.

THE GREAT IMITATORS

When the 19th-century naturalist Henry Walter Bates (1825–1892) was exploring the Brazilian rainforest, he noticed a curious fact: some clearwing butterflies appeared to have three pairs of walking legs, while others only had two. This consternating observation was, in fact, an example of perfect mimicry between Dismorphini and the brush-footed butterflies of the tribe Ithomiini.

Dismorphini go to such lengths to copy ithomiines because they themselves are not toxic while ithomiines are memorably unpalatable to predators. This type of mimicry is known as "Batesian," in honor of the person who first described it.

↓ *Dismorphia theucharila* (left) is a nearly perfect mimic of *Ithomia arduinna*, even though it is a pierid.

While its ithomiine model has four walking legs, *Dismorphia* has six, giving its identity away to keen-eyed observers.

→ The Provence Orange-tip, *Anthocharis euphenoides*, is one of about two dozen orange-tip butterflies that feed on mustards throughout the Holarctic region, mostly in warmer climates. The caterpillars are well-adapted to consuming these cruciferous plants, which are rich in glucosinolates.

BRIMSTONES AND CABBAGE WHITES

T he mimetic species of Pieridae described earlier in this chapter (see page 46) are more exceptions than the rule when it comes to this family. The typical shapes and colors of pierid butterflies are best exemplified by brimstones and cabbage whites.

LONG-LIVED BRIMSTONES

The caterpillar of the Common Brimstone, *Gonepteryx rhamni*, feeds on buckthorns (*Rhamnus*) and derives its Latin name from theirs. This butterfly has an interesting, well-researched biology: it overwinters as a reproductively inactive adult, then mates and lays eggs in the spring, which makes it one of the longest-lived species, with adults persisting for up to 10 months.

↓ The males of *Gonepteryx rhamni* are more lemon in color than the females, thanks to an additional pigment, and their wings reflect light in the UV spectrum that we cannot see.

HIDDEN PATTERNS

Males and females of *Gonepteryx* sport different color patterns, a result of males having two color pigments and females, only one. The bright, lemon color of the males changes slightly depending on the angle of incidence and shows up as a different pattern under ultraviolet light. This scattering of UV light in males, which is absent in females, results in the male butterflies having a slight iridescence. Males of *G. cleopatra*, found around the Mediterranean, are further adorned with a bright orange patch. See page 68 and Chapter 9, pages 114–116 for more on sexual dimorphism.

YOUR TYPICAL WHITE BUTTERFLY

The genus *Pieris* represents the typical, pure white butterfly—dashing erratically through fields, briefly stopping at flowers, all but consulting a pocket watch. Under this guise hide nearly 40 different species, which include the Large Cabbage White (*Pieris brassicae*), common to Europe; the Small Cabbage White (*P. rapae*) that is now found worldwide; and the Great Southern White (*Ascia monuste*) from the American tropics.

CABBAGE AND PEA LOVERS

The pierid family has a long history of coevolution between host plants and butterflies. The subfamily Coliadinae, which includes clouded yellows (*Colias*), small grass yellows (*Eurema*), and larger sulphurs (for example, *Phoebis*), frequently feed on legumes.

Whites belong to the Pierinae subfamily, which often relies on cruciferous host plants, but there are many exceptions. It is the most recently derived among the four subfamilies, diversifying approximately 50 Mya, and is now represented by six tribes, such as orange tips (Anthocharini), wood whites (Leptosiaini), and, of course, the whites (Pierini).

MASS MIGRATIONS

There are some notable exceptions to the sulphurs' affinity for legumes. For example, the spectacular mass migrations of the Lyside Sulphur, *Kricogonia lyside*, which can be observed in subtropical scrub habitats from Texas to Latin America, are precipitated by their

---

SULPHURS AND LEGUMES

Sulphur butterflies favor legume host plants, which can be found from suburban backyards to freezing stone outcrops in the Himalayas. The Cloudless Sulphur, *Phoebis sennae*, migrates with the seasons along the southeastern coast of the United States, keeping its eye out for its caterpillars' favorite, the partridge pea. Some species favor senna and cassia plants for oviposition, while *Colias* feed on alfalfa, vetches, pea-shrubs, and goat-thorns, among others. During the brief summers in the mountains of the Caucasus and Central Asia, one can find endemic *Colias* species zooming around at great speeds among these plants at high elevations, where the weather is punishingly cold for most of the year.

---

pursuit of lignum vitae (*Guaiacum officinale*), a shrubby plant in the caltrop family that's like mother's milk to their caterpillars.

Like the sulphurs, the Great Southern White, *Ascia monuste*, can travel high above the ground in such great assemblies that these mass migrations are easily detected by radar (see Chapter 8, pages 102–103 for more on migration). This species ranges from Florida to Argentina and is often lamented for being a frequent pest of cruciferous crops.

INTERTWINED WITH HUMANS

When humans migrate, they usually take their crops (and pests) with them. The history of the infamous Small Cabbage White, *Pieris rapae*, is quite interesting because it is closely tied to the spread and intermingling of human populations across the globe. Based on DNA analysis, *P. rapae* spread from its native range in the Mediterranean to northern Europe first. From there, it made its way to East Asia in the 18th century and then on to North America in the 19th century. By the first half of the 20th century, it had finally made it as far as Australia (via New Zealand).

↓ Caterpillars of the Small Cabbage White, *Pieris rapae*, derive defensive compounds from cabbage leaves called pinoresinols, which deters ants from attacking them.

BRUSH-FOOTED BUTTERFLIES (NYMPHALIDAE)

Nymphalidae, or "brush-footed" butterflies, form the largest and most diverse butterfly family, creating an umbrella for many unlikely relatives. They began to diversify 80 Mya, giving rise to very different subfamilies, such as the snout butterflies (Libytheinae) and milkweed and clearwing butterflies (Danainae). The snout butterflies form a very compact group of about a dozen species, some of which feed on sugarberry trees as caterpillars. As adults, they have uncharacteristically long palpi (for more about palpi, see Chapter 8, page 96), for which they received their common name.

TOXIC BEAUTIES

Milkweed butterflies (tribe Danaini) are characterized by their large size, slow flight, and affinity for milkweed host plants, such as *Asclepias* or *Cynanchum* (see Chapter 6, page 81). The smaller and more slender clearwing butterflies (tribe Ithomiini) often have partially or completely clear wings, making them almost invisible to predators. Others in this tribe have tiger-striped, orange-and-black wing patterns. These butterflies prefer to fly in the dark understory near their nightshade host plants and form mimicry rings throughout the Neotropics (see Chapter 9, page 117).

↓ A close-up of the head of the American Snout, *Libytheana carinenta* (left), versus that of the Painted Lady, *Vanessa cardui,* shows the long palpi for which the snouts got their name. Palpi have many sensilla and can detect the scent of ripe fruit.

→ The Neotropical Menapis Tigerwing, *Mechanitis menapis,* and the Southeast Asian Rice Paper Butterfly, *Idea leuconoe,* belong to the Danaine subfamily, in which bright colors serve as a reminder to predators of the butterflies' toxicity. Caterpillars also receive chemical protection from their host plants.

I SPOT AN EYESPOT

T he most recently derived lineage of Nymphalidae is the subfamily Satyrinae, which today unifies groups that used to belong to several different families. Most of these butterflies are small- to medium-sized with eyespots and an erratic, jumping flight pattern. Some break-off lineages, like the subtribe Euptychiina, underwent a tremendous radiation in South America, with an estimated 500 species that occur very locally in association with endemic bamboos, grasses, and occasionally lower plants, like mosses.

ELUSIVE NYMPHS

Some of the species that fly in the dark understory of South American rainforests, such as species of *Haetera* and *Cithaerias* in the tribe Haeterini, no longer have much need for color. Over time, they underwent such a drastic reduction in the number of wing scales that their wings became mostly transparent, with only hindwing eyespots left for adornment. This pattern fools predators into attacking only the wing margins, deflecting attacks from the vital head region. As a result, even if these practically invisible butterflies are spotted by a keen-eyed predator, they still have a good chance of escaping largely unharmed, with just a bit of their wing missing.

A FONDNESS FOR FERMENTATION

Satyrines are frequently referred to as "wood-nymphs," because, unlike many of their other nymphalid relatives, they can be found inside forests, flitting around just above the forest floor. This peculiar behavior can be tied to the diet of the adult butterflies— while they do occasionally feed on flowers, they are very fond of fermented fruit (hence their being named after the bibulous companions of Bacchus). And what better place to find some rotting fruit than on the forest floor?

OWL BUTTERFLIES

The Neotropical tribe Brassolini is comprised mostly of larger species that feed as caterpillars on the Zingiberales order of plants, from heliconias (such as wild plantains and lobster claws) to red ginger. Some of these species, such as the Forest Giant Owl, *Caligo eurilochus*, can even be a pest on cultivated bananas.
The wing patterns of *Caligo* butterflies actually do resemble avian eyes, and it is widely believed that these patterns serve as a warning to predators.

↓ Owl butterflies have eyespots that resemble those of birds of prey, thus supposedly dissuading attacks by smaller birds.

There are 17 genera recognized as members of Brassolini, from the medium-sized *Bia* to the giant *Caligo*, which reach over 6 in (16 cm) in wingspan—all found in rainforests, from Mexico to Argentina. They have become a mainstay of live butterfly exhibits because they are easy to rear, large, and eye-catching.

THE GREAT RADIATION
OF NYMPHALINAE

B etween the time that the snout butterflies and danaines (see page 52) evolved some 75 Mya and the appearance of satyrines 20 to 30 million years later, brush-footed butterflies underwent a rapid radiation. As a result, the family today is comprised of a dozen subfamilies, one of them being Nymphalinae.

FAMILIAR GARDEN BUTTERFLIES

Some nymphalid butterflies, such as peacocks and tortoiseshells (*Aglais*), buckeyes (*Junonia*), admirals and painted ladies (*Vanessa*), and Mourning Cloaks (*Nymphalis antiopa*), may be very familiar to people in Europe and the United States, where the number of butterfly

species per capita of butterfly enthusiasts is very low compared to more tropical countries. Other genera, such as daggerwings (*Marpesia*), which sport long hindwing tails and bright colors, are more familiar to people who live in South America or those visiting live butterfly exhibits, where these butterflies are frequently on display.

COMMUNAL SLUMBER

Like other animals, butterflies sometimes turn to "safety in numbers" as a strategy for survival. Nymphalids exhibit such clustering behavior, not only as eggs and caterpillars, but also as adults. For example, daggerwings and cracker butterflies (*Hamadryas*) form "roosts," aggregations of butterflies that come together to fold their wings up for the night. If one of the roosting butterflies is startled by a predator, they all fly up at once, allowing most of the participants to escape unscathed.

← The European Peacock, *Aglais io*, is one of the most beautiful and common butterflies found in Europe. These butterflies hibernate as adults and feed on nettles as caterpillars.

METALMARKS (RIODINIDAE)

Riodinidae is a family of small but very diverse butterflies that diverged over 70 Mya with their sister family Lycaenidae. They are mostly found in the Neotropics, with only a small representation in the Old World. The family is subdivided into seven tribes and around 130 genera, with more than ten times as many species.

FLEXIBLE GROUND PLAN

Riodinids are unified by similarities in wing veins and immature stages but are otherwise extremely variable. They range from species that have spectacular, shiny wings with long tails (like *Rhetus*) to amazingly accurate mimics of the clearwing butterflies (like *Stalachtis*). Some genera, such as *Calydna* or the numerous, far-ranging *Calephelis* are tiny—less than ³⁄₄ in (2 cm) in wingspan—with a fluttering flight that makes them easy to confuse with small, day-flying moths. By contrast, metalmarks of the genus *Eurybia* in the New World or of the genus *Dodona* in Asia are larger, achieving a 2¹⁄₂ in (6 cm) wingspan, and resemble miniature brush-footed butterflies in appearance and demeanor.

↙ Found in South America, the caterpillars of *Adelotypa annulifera* live on bamboo shoots and enjoy a symbiotic relationsip with ants that defend them in exchange for sugary secretions. The ants also don't attack visiting adult butterflies.

→ From the top: *Echydna punctata, Hyphilaria parthenis, Mesosemia loruhama.* These are just some examples of Neotropical riodinids. Male butterflies in this family frequently perch on leaves, monitoring their territory for females. This perching behavior, as well as the time of day at which it occurs, is species-specific.

59

FROM THE NEW WORLD AND BACK

There are two subfamilies of metalmarks recognized today: Riodininae and Nemeobiinae. They share a common evolutionary history, and both form mutualistic relationships with ants. Many riodinid caterpillars have special adaptations that allow them to signal to ants and reward them with sweet plant secretions. These ants, tamed by the caterpillars, remain aggressive toward other insects and vertebrates, thus providing these wily caterpillars with protection from predators.

OVER 90 PERCENT NEOTROPICAL

The riodinid family is thought to have originated in the Neotropics, and today over 1,300 species of riodinids can be found in the Latin American tropics, where they greatly diversified their shape, size, and natural history over millions of years. There are about 25 species that are found as far north as the United States, belonging mostly to the genera *Calephelis*, *Apodemia*, and *Emesis*. These species tend to have more conventional life histories than their jungle relatives.

SWEETENING THE DEAL

In the Neotropics, *Adelotypa* caterpillars feed on bamboo shoots, prompting the production of extrafloral nectar that draws ants and butterflies to these sweet droplets (see page 58). The ants do not display any aggression toward the butterflies, indicating that there must be some chemical and/or visual cues by which the ants accept them. On the other hand, *Pachythone xanthe* caterpillars, which feed on scale insects tended by Azteca ant shepherds, do not rely on the good graces of the ants. Instead, the caterpillars have developed a shield-like carapace for protection, and for added security, they also make use of sugar-secreting organs to bribe the ants.

OLD WORLD RIODINIDS

The other subfamily of metalmarks that is recognized today, Nemeobiinae, has a more complicated origin: it is mostly found in the Old World, where it diversified further by spreading to Asia, Africa, and Madagascar. Species like the spectacularly colored Malay Red Harlequin, *Paralaxita damajanti*, can be found from Borneo, Java, and the Malay Peninsula to southern Vietnam. The four species of the genus *Saribia* are found exclusively on Madagascar and sport intricate "false-head" patterns (see page 65 and Chapter 12, page 150). There is only one European member of the family, the Duke of Burgundy, *Hamearis lucina*, which looks like a miniature fritillary but has a caterpillar typical of riodinids that feeds on primulas. On the other side of the world, *Styx infernalis*, a highland species from Peru, resembles a pierid butterfly and, along with the Costa Rican Metalmark, *Corrachia leucoplaga*, is thought to have resulted from dispersal of an Old-World lineage to the Americas.

↓ *Styx infernalis* was at some point considered its own family of Lepidoptera but turned out to be a New World species of the mostly Old World subfamily of Nemeobiinae.

GOSSAMER-WINGED BUTTERFLIES (LYCAENIDAE)

The family Lycaenidae encompasses eight subfamilies. Among these are hairstreaks (Theclinae), blues (Polyommatinae), and coppers (Lycaeninae), which are more common across Europe, more numerous, and more recently evolved. The two ancestral lineages, comprising sunbeams (Curetinae) and harvesters and woolly legs (Miletinae), are mostly found in Southeast Asia, Indomalaya, Australia, and the Afrotropics.

A CATERPILLAR IN APHID'S CLOTHING

The subfamily Miletinae is less known than blues, hairstreaks, or coppers, but its species have remarkable life histories. For example, the caterpillar of the Harvester Butterfly, *Feniseca tarquinius*, the only miletine butterfly in North America, feeds not on plants but on woolly aphids, sometimes using their remains as camouflage. The evolution of this unusual behavior may be explained by the adult butterfly's affinity for the sweet secretions of these aphids and the major benefit gained through carnivory: transition from egg to pupa in just over a week.

↓ Carnivorous Harvester caterpillars feed on woolly aphids, which allows them to develop rapidly.

↓ The eggs of Lycaenidae, such as this one of the Common Blue, *Polyommatus icarus*, are sturdy and finely sculpted.

→ Blue butterflies in the genus *Polyommatus* include over 200 species whose males have bright upper-side coloration while the females are dark. They feed as caterpillars on various members of the pea family, such as the bird's-foot trefoil (*Lotus corniculatus*), vetches (*Vicia*), and crownvetches (*Securigera varia*).

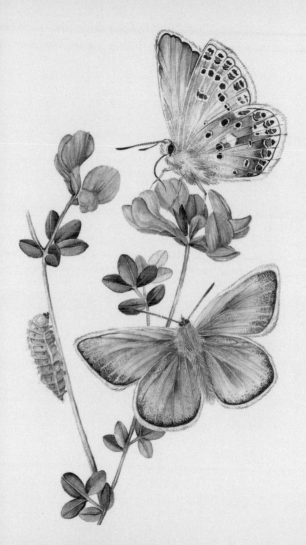

MASTERS OF ILLUSION

The subfamily Theclinae includes many butterflies that have a "false head": a pattern that converges on a spot at the tip of the hindwings, frequently accompanied by a little "tail." By moving their hindwings back and forth in a scissor-like motion, the butterfly gives the impression that the hindwing spot is its head and the wing tails are antennae. As a result, small ambush predators, such as jumping spiders and crab spiders, which lie in wait on flowers, attack these appendages, allowing the butterfly to get off lightly!

CONVERGENT EVOLUTION OF "FALSE HEADS"

The false-head pattern of Theclinae, beautifully exemplified by members of the genera *Arawacus*, *Thecla*, *Favonius*, *Atlides*, *Semanga*, and *Iraota*, can also be found in other families of Lepidoptera. It must have proven an extremely effective strategy for surviving lethal attacks by predators, as it has repeatedly evolved separately in families as diverse as swallowtails, nymphalids, skippers, and geometrid moths.

The same strategy is found in other groups of insects and even in vertebrates. For example, some fish, like the red drum (*Sciaenops ocellatus*) in the United States or the tailspot blenny (*Ecsenius stigmatura*) in the Western Pacific, have prominent eyespots on their tails to deflect predator attacks. But no group of animals has perfected this illusion better than the theclines!

TOXIC OFFSHOOTS

Some Theclinae species developed alternative defense strategies against predators by adopting toxic host plants. The Great Purple Hairstreak, *Atlides halesus*, supplements its false-head pattern with a penchant for feeding on mistletoe and a warning coloration to remind predators of this fact. Others, such as the Atala Butterfly from Florida and the Caribbean, as well as its Latin American relatives in the genus *Eumaeus*, sequester cycasin from their host plants, a compound that is highly toxic to vertebrates.

↑ *Arawacus separata*, Colombia. Lines converging on a tail and a spot attract even more attention to the "false head," which can be lost without harm to a fooled predator.

ENGINEERING ON SMALL SCALES

Many Eurasian theclines, like species in the genus *Chrysozephyrus*, have a spectacular iridescent coloration that is the result of nanostructures on the wing scales (see Chapter 9, page 110). Their eggs, which serve as the overwintering stage, must withstand harsh, cold conditions and inundations with water. The result is an elegant engineering solution executed on the microscale: spiky domes resembling coral but with a mathematical regularity in structure.

BUTTERFLIES ON FIRE

T he subfamily Lycaeninae are colloquially known as "coppers" in English, while in other languages their common name connotes "fire." Indeed, these butterflies look like ephemeral sparks flying across the verdant backdrop of meadows, appearing and disappearing as their wings open and close. This behavior signals to their mates and rivals and may also help to evade predators. Some coppers, such as *Athamantia* species in Central Asia, have little "tails" on their hindwings, creating false-head patterns like those of Theclinae.

COMMON COPPERS

The typical appearance of Lycaeninae is exemplified by the Scarce Copper, *Lycaena virgaureae* (once upon a time common in Europe), whose males have shiny, orange-red upper wing surfaces adorned with elegant black margins. Females have further adornment in the

form of black spots that stand out on this fiery orange backdrop. Another well-known member of the genus *Lycaena* (which includes over 80 species) is the Small Copper, *L. phlaeas*. This red-and-black butterfly is among the most widespread species in the world, ranging from Eurasia to northern Africa to North America.

CHEWING ON SORREL

Like many other members of the genus *Lycaena*, the caterpillars of the Small Copper feed on docks or sorrels (*Rumex*), with the eggs serving as the overwintering stage. Despite the fact that coppers feed on a relatively common host plant, their survival is not always guaranteed. The Large Copper, *L. dispar*, became extinct in the British Isles in the middle of the 19th century, when the damp habitats favored by the butterflies were converted to agricultural land.

← A Large Copper, Poland. This species is widely distributed throughout Europe, but unfortunately the subspecies native to Britain became extinct in the mid-19th century.

BLUES AREN'T ALWAYS BLUE

The subfamily Polyommatinae, colloquially known as "blues," comprises a number of common species whose males display sky-blue, sometimes iridescent, patterns on the upper surface of their wings. Female blues tend to be more modestly colored. In the European Adonis Blue, *Lysandra bellargus*, for example, they are a muted brown with orange spots.

SEXUAL DIMORPHISM AND SPECIATION

Rapidly evolving coloration and the stark difference between male and female wing patterns in this subfamily suggests that coloration plays a big role in sexual selection. Studies on the over 200 species in the genus *Polyommatus* found that the most closely related species actually had the largest variation in color. This suggests that these nuances in color are for the benefit of the female of the species, which relies on hues to recognize potential mates.

ENDANGERED MYRMECOPHILES

The endangered Miami Blue Butterfly, a subspecies of *Cyclargus thomasi*, feeds on nicker beans along the coast of the Florida Keys. This tiny butterfly has a non-obligatory association with ants, such as the Florida carpenter ant (pictured), which receive a sweet secretion from the caterpillars and in exchange do not attack them as they would otherwise. The Large Blue Butterfly, *Phengaris arion*, now rare in Europe, is not so sweet: its caterpillars parasitize *Myrmica* ant nests by chemically mimicking the ant queen, thereby tricking worker ants into feeding them at the expense of the ant colony.

VENTRAL WING PATTERNS

When blues are not in flight or mating, these butterflies tend to keep their wings closed, showing the world their less brilliant underside patterns, which often feature a series of concentric circles. Spotted patterns are helpful in visually disrupting the shape of a butterfly, assisting in camouflage. At close range, the marginal spots serve as deflectors of attacks from the head to the wing margin. Zebra blues (*Leptotes*), which include around 30 species predominantly found in Africa, sport not only spots but also little hindwing tails and a single bright spot, amplifying the "false-head" pattern similar to that of hairstreaks (see page 64).

CRYPTIC CATERPILLARS

Polyommatinae are mostly associated with herbaceous host plants, especially favoring nitrogen-rich legumes. Females frequently lay their eggs in the flowers themselves and on hatching, the caterpillars start their lives munching on succulent blooms, gaining both shelter and nutrition. As the caterpillars grow and pupate, they remain cryptic, camouflaged among petals and leaves. Since it's not possible to elude ants even with the best camouflage, these caterpillars produce sweet secretions to bribe these soldiers.

FOUR LIVES IN ONE

Metamorphosis is, in essence, about the division of labor. There are many benefits to having a complete cycle of metamorphosis with four stages of development: the embryo-housing eggs, the fast-growing caterpillar, the shape-shifting pupa, and the mate-seeking and egg-producing and -dispersing adult. This strategy gives an organism flexibility and grants each individual, population, and species a higher chance of survival.

ANY STAGE CAN DIAPAUSE

The egg can serve as a simple embryo incubator that endures no more than a week, or it can be an overwintering bunker that protects the fragile organism from harsh external conditions. In fact, any stage of development can enter diapause (a physiologically inactive stage) to deal with adverse circumstances, including the butterfly itself. The adult is usually only on the wing for a couple of weeks but can live up to 10 months, when necessary.

Every species has its own strategy. The Brown Hairstreak, *Thecla betulae*, glues its eggs to bird cherry twigs to see out the harsh winter. In many satyrines, such as the North American Common Wood-nymph, *Cercyonis pegala*, it is the newborn caterpillars that are the diapausing stage. In the Hungarian Glider, *Neptis rivularis*, the older caterpillar builds a winter shelter out of the host plant leaf. The Common Brimstone, *Gonepteryx rhamni*, and the European Peacock, *Aglais io*, slumber through the winter as mostly inactive adults, though the Peacock will show off its eyepots if disturbed, in an attempt to ward off predators.

↓ The Tawny Emperor, *Asterocampa clyton*, widespread through eastern North America, is attracted to tree sap and fermented fruit.

CATERPILLAR COMMUNES

Sometimes, the caterpillar is not only the growing stage but also the dispersing stage. When eggs are laid in large clusters, as is done by the Tawny Emperor, *Asterocampa clyton*, the caterpillars feed at first in a large group, collectively overcoming the plant's defenses. As they grow and need access to more leaves (hackberry for the Tawny Emperor), they strike out on their own and eventually pupate separately. In other cases, as in some tortoiseshell butterflies (*Aglais* species), the caterpillars remain together inside a huge silk nest spun over a patch of their nettle host plants and venture outside the nest only to feed and to pupate.

↓ Young caterpillars of the Tawny Emperor feed in large groups on trees of the genus *Celtis*. They hibernate as a third instar in smaller groups inside furled leaves.

EGGS: THE CRADLE

Butterflies can lay eggs singly or in batches, which determines whether the resulting caterpillars will start their lives alone or in a group. Typically, the ovipositing female will glue the egg to the host plant, although some satyrines, like the Spanish Gatekeeper, *Pyronia bathseba* and the Great Banded Grayling, *Brintesia circe*, have been observed parachuting their eggs into clumps of host plant grasses.

ARCHIT-EGG-TURE

Eggs can vary widely in shape, from perfect spheres to flattened or conical structures. They can be soft or hard, smooth or sculpted, or even spiny. The specific egg construction is determined by many factors, including the need to withstand inundations with water or survive long winters. An egg contains cytoplasm and a yolk, consisting of lipids, proteins, glycogen, and free carbohydrates. It is protected by a membrane and a shell, and air is delivered to the forming embryo through aeropyles.

↓ The egg of the Cabbage White butterfly is reinforced with vertical ribs. It is crowned with a rosette-shaped area and an opening called the micropyle, via which fertilization occurs before the egg is laid.

→ Life cycle of the Cabbage White, *Pieris rapae*. Eggs are laid singly and the larvae, vulnerable to many predators and parasitoids, are cryptically green. However, they are not entirely defenseless, producing a sticky secretion from the tips of their minute setae that is repellent to ants.

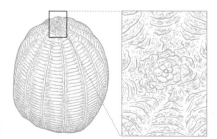

CATERPILLARS: THE BODY BUILDERS

With Absalom in *Alice in Wonderland* as a notable exception, caterpillars are rarely regarded with respect, even by those who love butterflies. They're viewed as voracious eating machines at best and crop-destroying pests at worst. Their ability to turn leaf matter into tissue is astounding—they can grow up to 1,000-fold in size in a single month. But few people look closely enough at these fascinating creatures to notice the many features that distinguish them from "worms."

FROM LEAF TO FRASS

In order to make use of the nutrients locked in plant fibers, a caterpillar must thoroughly masticate the incoming material. The caterpillar uses its labrum, a notched shield in front of the other mouthparts, to direct the leaf to the mandibles that chop it up. The shredded leaf is directed to the mouth by the labial and maxillary palps. After passing through the digestive system (see Chapter 7, pages 84–85), the waste is formed into barrel-shaped frass pellets, roughly the size of the caterpillar's head, and is expelled, sometimes purposefully catapulted away to avoid detection by predators (see Chapter 12, page 146).

CATERPILLAR DEFENSES

Caterpillars have a variety of strategies when it comes to self-defense. Many are colorful, hinting that they are chemically defended (i.e., toxic), thanks to the plant compounds they sequester from their diet. Some species, such as swallowtails, have osmeteria, glands that release an unpleasant odor when the caterpillar feels threatened. Others, such as lycaenids and metalmarks, rely on *quid pro quo*, producing sweet secretions for ants in exchange for protection (see Chapter 5, page 68).

← The caterpillar of the Plain Nawab butterfly, *Polyura hebe*, from Southeast Asia, has head horns as a defense against enemies.

SPINNING SILK

Beneath the mouthparts, one can spot a spinneret: an elongated organ that releases silk. A pair of silk glands spans the length of the caterpillar body and produces a liquid mixture of proteins that is extruded as strong, thin fibers. While silk is primarily associated with the way in which moth caterpillars make their cocoons before pupating, nearly all Lepidoptera caterpillars produce this useful substance. Butterfly caterpillars use it for laying pads, or anchor points, on leaves while feeding, constructing shelters out of leaves, and firmly attaching a chrysalis prior to pupation. When caterpillars outgrow their skin, they secure themselves with silk while they molt.

In addition to these active forms of defense, most caterpillars also try to keep a low profile by utilizing camouflage and living in inconspicuous shelters. If they are discovered, they can try to ward off predators and parasitoids with their head horns. If they're grabbed, they'll hold on for dear life to their leaf using crochets (sharp hooks located on their prolegs). Some just go limp at the first sign of danger, tumble off the plant, and play dead, hoping to be obscured from view by the undergrowth.

SENSING THE WORLD

Caterpillars have six pairs of eyes known as stemmata (see Chapter 8, page 98), short antennae with which they sense smell and physical touch, and taste receptors located in the epipharynx. The antennae are located below the stemmata and point downward, with a bulbous sensilla located at the tip. A caterpillar's sense of smell works best at short ranges and provides crucial information about specific plant volatiles, helping them choose the right leaves.

CHRYSALIDES: THE HIDE-AND-SEEKERS

Before pupating, the caterpillar sheds excess food from its gut, finds a place that is secure enough for the future chrysalis, and becomes a prepupa, attaching itself to the substrate with silk. At this point, its gut is emptied, its fat stores are maximized, and hormones abound, transforming embryonic wings, gonads, and other organs into their adult counterparts.

AVOIDING ATTENTION

Many caterpillars will leave their childhood homes (i.e., their host plants) to pupate, because it is the first place that parasitoids and predators will look for them, and they will be at their most vulnerable as chrysalides. The caterpillars of the Palearctic Large Tortoiseshell, *Nymphalis polychloros*, feed gregariously, stripping whole branches of elms or willows, but when they mature, they crawl away from the feeding site and pupate on the trunk or away from the tree entirely. Monarchs and pipevine swallowtails in North America are frequently found pupating on artificial structures, such as road signs and buildings. Many skippers pupate inside rolled leaves, sometimes making a silk cocoon. In some ancestral butterfly lineages (see Chapter 3, page 36), caterpillars bury themselves in the soil (*Baronia*) or under rocks (*Parnassius*) before pupating.

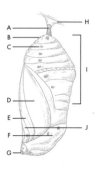

← On the surface of the chrysalis, one can see the body parts of the future adult butterfly. (A) cremaster, (B) genitalia, (C) spiracle, (D) wing, (E) antenna, (F) thorax, (G) head, (H) silk pad, (I) abdomen, (J) legs.

→ Within the context of their environment, most chrysalides are cryptic, but there are some exceptions. (A) Riodinidae (*Melanis*), (B) Hedylidae (*Macrosoma*), (C) Pieridae (*Pieris*), (D) Lycaenidae (*Eumaeus*), (E) Hesperiidae (*Tagiades*), (F) Papilionidae (*Troides*), (G) Nymphalidae (*Euphydryas*).

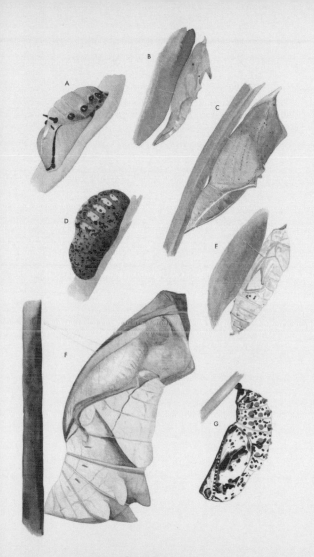

THE NATURAL ENEMIES

Thanks to numerous observations and experiments, we know that predation by birds drives the evolution of butterfly color patterns and of their chemical defense. But caterpillars have an even greater host of enemies, for which they have developed additional defenses.

FOOD FOR VERTEBRATES

Caterpillars are an important source of food for young nestling birds. If you have ever observed how, in the early morning, small warblers inspect every inch of a tree branch when foraging, you realize how hard it would be for a caterpillar on that branch to escape detection.

WASP PARASITOIDS AND PREDATORS

Despite the obvious role of bird predation on caterpillars, it is invertebrates that apply the largest selective pressure. These predators and parasitoids are unimaginably numerous. It is estimated that only 5–12 percent of approximately 800,000 parasitoid species have been described by science, and they have developed an array of weapons specifically targeted at a narrow range of prey, making them extremely effective.

PREDATION BY ANOLES

Lizards are opportunistic hunters of caterpillars. The green anole lizard, which was introduced to Japan, has been linked to the extinction of the Ogasawara Holly Blue, *Celastrina ogasawaraensis*. In the United States, it's out of the frying pan and into the fire: brown anoles, introduced from the Caribbean, are outcompeting the native green anoles and frequently becoming predators, in turn, of butterflies and their caterpillars.

Female parasitoids can detect the volatile chemicals produced by a host plant under a caterpillar's attack and home in on their victims. They have syringe-like ovipositors that they use to inject eggs into the caterpillars, often temporarily paralyzing them with venom to prevent them thrashing around. Potter and thread-waisted wasps also paralyze caterpillars, but permanently, using them as "canned food" for their larvae. The paralyzed caterpillar is sealed with the wasp's egg in a pot-like mud chamber, in which the wasp larva will develop.

↑ Ichneumonid parasitoid wasp laying an egg inside a caterpillar. Its larva develops within the hemolymph while the caterpillar continues to grow, ultimately killing its host.

HITCHING A RIDE

It is not only caterpillars and chrysalids that can come under parasitoid attack. Tiny trichogrammatid wasps (*Xenufens* species) will hitch a ride on the huge wings of the female Forest Giant Owl Butterfly, *Caligo eurilochus*, until it lays its eggs. The wasps will then dismount and lay their eggs inside the butterfly eggs. The wasps are so tiny that they can fully develop into an adult inside a single butterfly egg.

DEADLY AND INTERESTING PATHOGENS

Viruses, bacteria, and fungi also take their toll on caterpillars. If you have ever seen a caterpillar hanging from a tree branch, black and lifeless, as if liquified on the inside, it has probably been a victim of a baculoviral infection. Fungal spores lay dormant on host plant leaves and when ingested by a caterpillar, grow invisibly inside them, eventually turning them into mycelium-stuffed mummies.

Not all pathogenic bacteria are fatal to caterpillars. At the turn of the 20th century, scientists noticed that some African butterflies in the genus *Acraea* had a sex ratio greatly skewed toward females, and it was discovered that bacteria called *Wolbachia* are responsible. Genetically male caterpillars would develop into female adult butterflies, but an antibiotic treatment administered to the infected caterpillars would restore the 1:1 sex ratio.

BUTTERFLY AND PLANT COEVOLUTION

While caterpillars are voracious eaters, they are not indiscriminate. Few caterpillars are polyphagous (feed on multiple host plants) and most will only consume a narrow range of plants. Sometimes, their diet is restricted to a single plant species. Their physical development and their evolution is inexorably tied to their chosen plants.

TASTE RUNS IN THE FAMILY

Female butterflies are very particular about the plants on which they lay their eggs, because each plant has its own set of chemical and other defenses that young caterpillars will have to overcome. Taxonomic groups of butterflies that are united by a common evolutionary history frequently utilize similar host plants. For example, among swallowtails, most Apollo butterflies (*Parnassius*) feed on stonecrops and troidines feed on pipevines (see Chapter 3, pages 36–39). Among pierids, many feed on legumes or cruciferous plants (see Chapter 4, page 50), although *Delias* species depart from the norm with their choice of mistletoe (see Chapter 4, page 51).

GENETICS OF DETOX

Using genomic research, scientists are just beginning to understand the evolutionary pathways that have enabled butterflies and caterpillars to sequester toxic plant compounds for their own use. For example, differences in the ATPα gene have been shown to account for the Monarch's resistance to cardiac glycosides, compounds found in milkweeds.

This was determined by first sequencing the genomes of several different insect orders that are all able to feed on milkweeds with impunity and identifying the DNA sequences they have in common. Then, to test that these sequences were actually tied to milkweed immunity, the genomes of fruit flies were altered to introduce these DNA sequences. Remarkably, these "monarch fruit flies" gained resistance to milkweed chemicals, establishing the genetic link.

← Aposematically colored caterpillars feeding on a mistletoe and the adult of the Painted Jezebel, *Delias hyparete*, a highly variable butterfly found in Southeast Asia.

RESPIRATION

Insects don't have a four-chambered heart that pumps red blood through arteries, but they do have a circulatory system that performs a similar function. Whereas the human body needs minute capillaries extending to each extremity and weaving through every corner of an organ to deliver oxygen and whisk away carbon dioxide, an insect's hemolymph washes the organs freely. It is pumped throughout the body by a long, tube-shaped organ called the dorsal vessel that runs along the back of the insect. Just like blood, hemolymph carries nutrients, waste products, hormones, and immune cells throughout the organism.

GAS EXCHANGE

Respiration in insects is accomplished through the tracheal system: a network of branching, incredibly thin, chitinous tubes that deliver oxygen throughout the body. The nine openings of the trachea are located at the spiracles along the thorax (two) and abdomen (seven) of the adult butterfly. In the caterpillar, the distribution of spiracles changes to one and eight, respectively.

Gas exchange in the tracheoles (the terminal branches of each tree-like trachea) is far from passive diffusion, as was long assumed, but is, in fact, assisted by the contraction of muscles and the pumping of hemolymph. The interplay between pressure gradients in the tracheoles and the opening and closing of spiracles creates a temporarily closed air circulation system, which allows for oxygen to be pushed further into the tissues.

← Diagram showing branching trachea and tracheoles within a butterfly chrysalis.

These branching tubes penetrate wings and other tissues, supplying them with oxygen.

AIR IN THE SAILS

Butterfly wings are living tissue, which means they also require oxygen. The trachea and the hemolymph are enclosed by long, cuticular, branching tubes—the wing veins—along with an epidermal layer and nerves. The hemolymph is pumped by expanding and contracting muscles (the "wing hearts"), which are located in the thorax. When the local volume of hemolymph increases, it exerts pressure on the elastic trachea, pushing out the stale air inside them. When the hemolymph is pumped out and its volume decreases, the negative pressure in the trachea causes fresh air to be pulled into the wings.

↑ Morpho butterfly wing shows veins containing trachea and nerves. The pumping of hemolymph via the veins causes wings to expand when the butterfly exits the chrysalis.

FEEDING AND DIGESTION

While the larval and adult stages have divergent diets due to their respective energy demands, there are some common themes to their digestive systems. Both have a foregut, midgut, and hindgut, which consists of the ileum and rectum. Malpighian tubules collect dissolved by-products of metabolism from tissues and direct them to the posterior rectum, where water and salts are absorbed and the rest is excreted.

DIGESTION IN THE MIDGUT

The gut contains numerous bacteria and fungi that assist in digesting cellulose, synthesizing necessary vitamins, and defending against pathogens and insecticidal plant chemicals in exchange for a stable environment in which they're able to thrive. The midgut is where most of the digestion happens, including the breakdown of plant matter and bacteria from which nutrients are obtained. It is lined with a peritrophic matrix, which is responsible for immune defense and prevents bacteria and fungi from coming into direct contact with the epithelium and causing infection.

TRANSFORMATIONS IN THE DIGESTIVE TRACT

The digestive system shrinks in size over the course of metamorphosis. The foregut, where chewed but not yet digested food is stored in the caterpillar, becomes much thinner and longer during the pupal stage. The pupa also develops a "crop," a sac connected to the foregut, which is full of liquid in the pupa but empty in the adult. It is thought that its function is to store the extra fluid needed to expand the butterfly's wings upon emergence.

CHANGES IN DIET THROUGH LIFE

The transformation of the digestive system during metamorphosis is related to the dramatic change in the type of food consumed by the adult. Caterpillars consume large volumes of solid plant matter, while most butterflies feed on liquid sugars and amino acids found either in nectar or in fermented fruit—high-energy food needed to sustain flight. They pump this liquid food through a long, thin proboscis into their foregut, eliminating the need for chewing.

Longwing butterflies (*Heliconius*) are an interesting case because, in addition to nectar, they also feed on the pollen that gets stuck to their proboscis. However, even in this case, the solid food is first digested externally by the release of enzymes that break the pollen proteins into amino acids, which are then ingested with liquid. The addition of pollen to the diet is thought to be correlated with the notable longevity of these butterflies.

↓ *Heliconius* digest pollen that sticks to their proboscis. Imbibing amino acids from this food source allows them to live longer than other butterflies without hibernation.

EXCRETION

T he Malpighian tubules form a set of thin passageways that penetrate the body, collecting the by-products of metabolism from the hemolymph and delivering them to the hindgut where some are excreted and some are reabsorbed. Since caterpillars and butterflies feed on moist foods, the diluted metabolites, as well as sodium and potassium, can travel directly to the gut.

RECYCLING SALTS

Caterpillars need to maintain a very high pH in their midgut to grow as fast as they do, since alkaline conditions are more favorable for symbiotic microorganisms and for the digestion and absorption of food. To sustain this environment, it's necessary for caterpillars to have the ability to reabsorb the potassium excreted from tissues during the metabolic process. This is accomplished by tubules in the hindgut that form the cryptonephridial system, a cluster of vessels within the inner gut wall. In liquid-feeding adult butterflies, the system is remodeled to allow for faster fluid excretion.

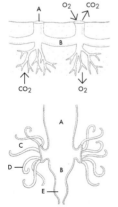

← Through openings along the body called spiracles (A), oxygenated air is supplied to tissues via tubes called trachea (B) that branch into progressively smaller tubes.

← Malpighian tubules (D) connect to intestines where the midgut (A) joins the hindgut (B). Together, these systems get rid of metabolic wastes from the gut and hemolymph (C) via the rectum (E).

→ Systems of the adult butterfly: (A) nervous, (B) digestive, (C) Malpighian tubule, (D) circulatory, (E) reproductive system. Most of the organs exist in the caterpillar as well, but some undergo restructuring during metamorphosis (arrows show the flow of hemolymph [see page 83]). The respiratory system (trachea) is not shown here (see page 82).

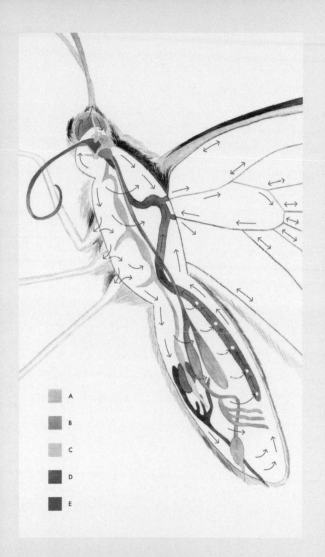

A

B

C

D

E

HORMONES

Metamorphosis is regulated through the waxing and waning of hormones. Most adult butterfly organs are already present in some form in the full-grown caterpillar, but they undergo significant restructuring over the course of metamorphosis. Wings and male gonads develop more gradually as the caterpillar matures, while the female reproductive organs, proboscis, and compound eyes only form inside the chrysalis.

↓ Meadow Brown females, *Maniola jurtina*, can fall into reproductive diapause during the summer heat.

There are three primary hormones that control the essential functions in insect metamorphosis: the brain hormone, the juvenile hormone, and the molting hormone, known as ecdysone. To date, over 30 insect hormones have been identified.

MONARCH MIGRATION AND THE JUVENILE HORMONE

Upon emergence from the chrysalis in the fall, the Monarch butterfly (see Chapter 8, pages 102–103) often has low levels of juvenile hormone, so instead of producing eggs and sperm, it will exhibit very different behavior: it starts migrating to its overwintering site, accumulating fat in the process. These butterflies live much longer. Males and females settle together, sometimes by the millions, to overwinter, but they do not mate until the spring, when the juvenile hormone levels rise in response to the lengthening days and warmer temperatures. Females of this generation then begin producing and laying eggs, giving rise to the next three generations that recolonize the vast territories abandoned by the monarchs in the fall.

HORMONE DISCOVERY

In the early 20th century, the existence of the brain hormone was proposed based on experiments conducted on Gypsy Moth (*Lymantria dispar*) caterpillars, in which their brains were extracted at different stages of development. The important role of the corpora allata in metamorphosis was discovered in the 1930s through studies on silkworms. In the ensuing two decades, three species of giant silk moths were used as model organisms to understand the role of the juvenile hormone via surgical experiments and chemical analysis. Today, at least four different variants of the juvenile hormone have been discovered in insects, some unique to Lepidoptera.

BRAIN AND JUVENILE HORMONES

The brain hormone primarily regulates the production of ecdysteroid hormones by the prothoracic gland, but evidence suggests that it may also play a role in egg production. When it is time to molt, the prothoracic gland, located in the first segment of the thorax, releases ecdysone under the influence of the brain hormone. Because of this connection, the brain hormone is also known as the prothoracicotropic hormone (PTTH).

Additional hormones regulate the shedding of the old cuticle, trachea, and head capsule during molting, including the old mandibles that have been worn out by constant chewing.

JUVENILE HORMONE AND METAMORPHOSIS

Levels of juvenile hormone (produced in the corpora allata) and ecdysteroid hormone dictate whether the caterpillar pupates or molts into the next instar. Pupation is inhibited when juvenile hormone levels are high, but as soon as they drop, the caterpillar metamorphoses into a chrysalis. In an adult butterfly, the juvenile hormone controls mating and egg-laying behaviors. For example, the reproductive behavior of the Meadow Brown, *Maniola jurtina*, described on the following page, and that of the overwintering Monarch, *Danaus plexippus*, is regulated by this hormone.

NUPTIAL GIFTS AND MATING

E xperiments with the Green-veined White, *Pieris napi*, demonstrate that females can be "promiscuous" on purpose, obtaining valuable nutrients from each coupling, including those for future eggs. Males transfer 15–23 percent of their body mass to females during mating and then need to recharge their mating potential by feeding and puddling (obtaining nutrients from wet surfaces, such as river banks). A female can mate up to five times to maximize egg size and fecundity.

A COOL HEAD IN THE HOT SUMMER

During hot summers in Italy, male Meadow Browns, *Maniola jurtina*, hatch from their chrysalides a few days before the females. Both sexes are ready to mate within a day or two. After mating, females go into a reproductive diapause to avoid the mid-summer heat—there is little incentive to form or lay eggs since there is no fresh food available for their caterpillars. So, they hide, biding their time until the cooler, wetter autumnal months, and only then form and lay their eggs to provide their offspring with the most favorable conditions.

↓ Male genitalia: (A) tegumen, (B) anus, (C) valva, (D) penis, (E) ejaculatory duct, (F) anal tube. Genitalia are a ring-shaped structure (tegumen is its dorsal part) to which moving parts are attached.

↘ Female genitalia: (A) ovary, (B) anal tube, (C) oviduct, (D) spermatheca, (E) accessory gland, (F) vagina, (G) bulla seminalis, (H) bursa copulatrix (this stores the spermatophores).

→ Cairns Birdwing, *Ornithoptera euphorion*, Australia. Males of this species have been known to stay with the females for the entire day after mating, warding off competitors.

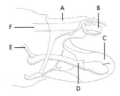

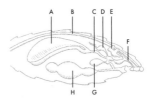

SPERM AND EGG FORMATION

Female butterflies seldom live long once their eggs have been laid, so their reproductive system is set up to make the moment count. Some butterflies have eggs mostly formed and ready to lay upon emergence from pupae, while others continue forming them over a longer period of time. Eggs form from germ cells and as they pass through one of the four ovarian tubes, they are furnished with nutrients and an outer shell called the "chorion."

STORING THE SPERM

Most butterflies produce eggs gradually and lay them singly or in small batches over a period of two to three weeks. To be able to fertilize the newly formed eggs, the female needs a way to store the sperm it has received from the male as part of a package called the "spermatophore," which also contains useful nutrients. Spermatophores are stored in the female's bursa copulatrix, where they can be broken down and used as needed. By counting the number of spermatophores in the bursa copulatrix of a dead specimen, scientists can tell how many times the female has mated.

TWO TYPES OF SPERM

There are two types of sperm produced by the male: the eupyrene, which carry genetic material in their nucleus, and the apyrene, which have no nucleus. The function of the second type is still being debated more than a century after its discovery. Apyrene sperm migrate to the spermatheca first, possibly signaling to the female that they no longer need to seek out a mate but should instead be laying eggs. There is also evidence that these sperm may play a role in sperm competition between rival males by overwhelming competitors' sperm from any previous couplings.

FERTILIZATION AND EGG LAYING

Once the spermatophore is broken down, the sperm is passed into another, much smaller sac called the spermatheca. From there, sperm cells swim through a duct into the vagina and fertilize the eggs, entering the egg through a micropyle opening, following which the egg membrane normally blocks further fertilization. In the laying process, eggs are enclosed in tough chorion and are glued to a leaf or branch with sticky secretions from the colleterial gland. Experiments on a satyrnine, *Bicyclus anynana*, showed that larger (but fewer) eggs are laid at lower temperatures, providing developing embryos with better nutrition, but resulting in fewer offspring.

↑ The Map Butterfly, *Araschnia levana*, lays eggs in strings on the underside of stinging nettles, mimicking the flowers of their host plant and allowing larvae to feed in groups.

EVOLUTION OF DIURNAL BEHAVIOR

Butterflies are known for their diurnal habits and are most frequently seen flying in broad daylight. However, there are a great variety of light-related behaviors, from butterflies that fly in the darkness of the rainforest, close to the forest floor, to species that prefer dusk. Butterflies have evolved corresponding adaptations for such behaviors, such as preferentially discerning certain colors or having better night vision.

Butterflies have an extensively developed sense of smell and are attracted to a variety of volatile substances. In addition to compounds produced by flowers and host plants, they are also attracted to the scents of ammonia and other products of fermentation in rotten fruit and the scent of urine, rotting meat, and carnivore dung, which promise accessible nitrogen.

THE FLOWER BUFFET

Butterflies serve as pollinators, so flowers are incentivized to adapt their presentation to be as eye-catching as possible to these nectar-seekers. Flowers tailored to diurnal pollinators tend to be large and colorful, often red or yellow, while those that cater to crepuscular species are small and white. Many plants have evolved to bloom and produce nectar within specific time windows, as well as manufacture special fragrances, with the goal of attracting particular butterflies that will pollinate and enhance reproductive success for those plants.

NIGHT VISION VERSUS COLOR VISION

Registering a broader spectrum of light generally comes at the expense of seeing well in the dark. Both crepuscular owl butterflies (*Caligo*) and their diurnal relatives, the morphos, feed on fermenting fruit on the forest floor. Because owl butterflies do this at dusk with very little light, their eyes have evolved structural adaptations, such as enlarged facets, to enable them to see better in these conditions. Morphos, on the other hand, have preserved a better ability to distinguish colors.

← The Old World Swallowtail, *Papilio machaon*. In addition to visual pigments in the eyes called opsins that are responsible for color vision, other multipurpose proteins are also involved in visual processing.

RECEPTORS AND ODORS

You often see a female butterfly flying through a forest or meadow searching for a suitable host plant on which to lay its eggs. Instead of heading straight to the right plant, the butterfly lands on numerous wrong ones along the way, but this temporary alightment has a purpose—her feet have taste receptors, so this is actually a quick way to evaluate candidates.

WHAT PERKS UP BUTTERFLY ANTENNAE?

Scientists can study the chemistry of insect–plant interactions using gas chromatography-electroantennographic detectors (GC-EAD). Volatiles or chemical compounds are collected from the plants of interest and run through a mass spectrometer, which separates compounds according to their relative masses and ejects them onto the antenna, leg, or other body part of the butterfly. These are connected to an electrode, which registers the electrical signal from the neurons "firing" in response to compounds that the butterfly is able to detect. The numerous plant compounds to which the butterfly reacts can be further pared down by using existing databases.

↓ The butterfly senses include olfaction (sense of smell), for which antennae (A) are primarily responsible, but palpi (B) are also implicated. Color vision is achieved via compound eyes (C), and there are numerous taste receptors on the proboscis (D).

→ *Colotis amata*, a pierid found in Africa and Asia, laying eggs on the leaf of a toothbrush tree (*Salvadora persica*). All butterflies have receptors located on their legs, which helps females recognize the correct host plants. Similarly, a butterfly can taste food with its legs. For example, if it touches a sugary syrup with its front legs, the hungry butterfly will instinctively uncoil its proboscis.

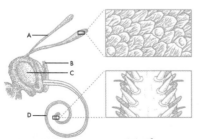

EYES AND VISION

The eye of a butterfly consists of many facets called ommatidia, which combine lots of individual snapshots of the surroundings into one image through processing in the brain. Butterflies can see a broader spectrum of light than humans, and they have three types of specialized receptors that register ultraviolet (300–400 nm), visible blue light (400–500 nm), and the longer wavelengths in the visible and infrared regimes (above 500 nm).

SPECIES-SPECIFIC VISUAL STRUCTURES

Butterfly visual structures have evolved differently for different species to adapt to their specific needs and environments. Apposition eyes, where each ommatidium is well-screened from the light entering the neighboring ones by light-absorbing pigments, are better for seeing in bright light while superposition eyes are optimized for seeing in dimmer conditions because the light is purposefully distributed across adjacent ommatidia.

Many butterflies are drawn to red and yellow flowers when searching for nectar. In the brush-footed butterflies (Nymphalidae), red-screening pigments can improve sensitivity to red color in some

---

CATERPILLAR VISION

Caterpillars use vision when crawling around in search of food and for detecting predators. They have six pairs of eyes (called stemmata) arranged in a circle on the sides of the head. Their eyes are much simpler than those of adults, with each stemma being a simple lens akin to a single ommatidium. However, these simple structures can work pretty well, especially if each eye performs a different function, as demonstrated in a study of processionary caterpillars of two prominent moth species: *Ochrogaster lunifer*, and *Thaumetopoea pityocampa*.

receptors. In certain butterfly species, such as the Clouded Yellow, *Colias erate* (Pieridae), males see better in some ranges of the light spectrum than females due to the presence of different pigments in some of their ommatidia, which are thought to play a role in mating behavior.

REMARKABLE OPSINS

Light, after passing though the cornea, is focused by the lens onto the rhabdom, where the color-sensitive pigments (chromophores attached to protein groups called opsins) are located inside photoreceptor cells. Ommatidia can differ from each other in their ability to detect different colors within the same eye, and inside an ommatidium there can be individual cells that are only sensitive to a very narrow range of wavelengths. For example, in the Monarch, *Danaus plexippus*, there are some cells that only process information from polarized UV light, which becomes important for navigation (see page 102). Via duplication of certain opsin-coding genes, the Atala Butterfly, *Eumaeus atala*, has adapted to see red colors better, which is perhaps an important trait in locating mates and flowers while avoiding host plants that are already populated with Atala caterpillars (also red).

EARS ON THE WINGS

On the ventral surface at the base of the forewing of many nymphalid butterflies, buried underneath scales, one can find a specialized hearing structure called the Vogel organ. Although the Vogel organ was first described more than 100 years ago, its function has been tested in laboratory experiments on only a handful of species. In the Common Wood-nymph, *Cercyonis pegala*, the enlarged vein of the forewing also serves as part of a hearing organ, and it is possible that such an adaptation is widespread among butterflies.

DIFFERENT PREDATORS, DIFFERENT CUES

When one compares the sound frequencies to which butterflies and moths respond, it appears that they are listening for different predators. The attunement of moths to the high-frequency ultrasonic signals of bats has been well-documented. The crepuscular Neotropical White-spotted Satyr, *Manataria maculata*, can also detect higher-frequency sounds, perhaps on the lookout for bats, but most diurnal butterflies hear in a much lower range, most likely attuned to diurnal predators like birds.

→ The Vogel organ of the Blue Morpho, *Morpho peleides*, is innervated by nerves (A) that lead to the mechanoreceptors (B), which are covered by the tympanic membrane

(C). The organ is most sensitive to low frequencies at a noise level of 58 dB (like the sound of your fridge), so, unlike in moths, it is not normally used to detect ultrasound.

→ The Common Wood-nymph, *Cercyonis pegala*, has a hollow coastal vein in the forewing that serves as an amplifier of low frequency sounds. At the base of the forewing is a tympanal ear, which suffers a loss in sensitivity if the vein is removed.

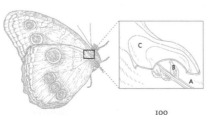

ORIENTATION AND MIGRATION

Many butterflies migrate with the seasons. The Painted Lady, *Vanessa cardui*, populates the Northern Hemisphere all the way to the Arctic Circle during the summer and then migrates back to its breeding grounds in the fall. Among all migratory butterflies, the Monarch, *Danaus plexippus*, has garnered the most attention for its spectacular fall migration. These large, brightly colored butterflies can be observed, sometimes in great numbers, making their way down to Mexico all the way from Canada.

NAVIGATION

Monarchs either fly to Mexico from the eastern part of North America or to roosting sites along the Pacific coast of the United States when they breed west of the Rockies. They also migrate from the Australian coast to sites near Sydney and Adelaide. On their long journey to Mexico, they can visually assess the angle of the sun by detecting polarized sunlight, especially in the UV spectrum, while their circadian clock tracks the length of the day. Even when it is overcast, they can still orient themselves thanks to a backup compass located in their antennae that relies on Earth's magnetic field.

THE GENETIC BASIS OF MIGRATION

When the caterpillars of migrating monarch butterflies mature, the daylight hours are getting shorter, which triggers the expression of genes responsible for migration. The resulting monarchs have differently shaped wings and different metabolisms. They are born in reproductive diapause, so instead of mating, they feed and fly south, accumulating fat that will be crucial for surviving the winter. Over 16,000 protein-encoding genes have been identified in the Monarch genome, with over 100 of these differing between migratory and non-migratory butterflies.

ARRIVING TOGETHER

Monarchs pace themselves while flying south/southwest in the fall, so as to arrive in Mexico in early November, with late arrivals trickling in by December. By marking and releasing monarch butterflies, scientists have figured out that they pace themselves by keeping the sun angle at solar noon constant (at around 57°). Along the way, monarchs form temporary roosts and decide to fly or not depending on the weather, so each individual's journey is unique. In this sense, their migration is completely different to that of birds, which rely on social behavior to stay together.

HEADING BACK NORTH

As the days become longer and warmer, in February to March, monarch butterflies in the Mexican overwintering colonies begin to mate and fly north (see also Chapter 7, page 88). As they find sprouting milkweeds, they start laying eggs and the next two generations continue the process of repopulating the North American continent, exploiting milkweed resources. This behavior enables a tropical butterfly, which is unable to withstand the harshness of northern winters, to still take advantage of abundant breeding grounds in northern climates.

STUDYING MIGRATION

Much of what we know about navigation and the genetic mechanism of insect migration is thanks to research on monarch butterflies. Techniques employed over the years to build a better understanding of migration have included mark-and-recapture programs, where identifying tags are placed on the wings; flight simulators, where a butterfly is suspended in an observation chamber equipped with a camera; transporting and releasing butterflies and observing their subsequent behavior; and sequencing and manipulating genomes and otherwise altering individual butterflies. Even the brain of monarchs has been mapped, documenting the minutiae of its complex structure and function.

↓ In the overwintering Monarch colonies in Mexico, before beginning to disperse, males chase females and tackle them to the ground trying to mate.

FLIGHT AND PERCHING

While their flight may seem elegantly deliberate and well-controlled, butterflies have only indirect control of their wings. When the muscles of the thorax contract, the resulting pulse produces several wing flaps. The point of wing attachment to the thorax, where these muscles are located, has numerous sclerites, which resemble our wrist bones in complexity and number. From the base of the wing, many branching veins radiate, giving it strength and flexibility. Wing veins contain trachea and hemolymph and are equipped with sensory organs.

WINGS OF MANY SHAPES AND SIZES

Different wing shapes and bodies mean different flight abilities and habits in butterflies. Some members of the Swallowtail family (Papilionidae), such as birdwing butterflies, have large, triangular forewings and small hindwings and propel themselves with great speed through the air using the powerful flight muscles of their long and wide thorax. Skipper butterflies (Hesperiidae) resemble miniature birdwings in this respect, but perhaps due to their small size, they are unable to glide, instead dropping downward after each wing stroke. Their small size and powerful muscles allow them to react faster than any other butterfly to a threat when they are at rest.

In contrast, the Paper Kite Butterfly, *Idea leuconoe*, and longwings of the genus *Heliconius* (both in the family Nymphalidae) have relatively small, short thoraxes, indicative of weaker flight muscles, but long, oval-shaped wings that allow them to effortlessly hover in the air over flowers. They are less concerned with escaping predators since they are toxic, so they "educate" birds instead of avoiding them by giving them unpleasant gustatory experiences to recall along with their wing patterns.

FLIGHT BEHAVIORS

Butterflies can often be seen sunning themselves on a cold day, lazily opening and closing their wings. They quickly warm up from the sun's rays, but when they take flight, the cold air cools them quickly, so they must land again to repeat the exercise regularly. Males of many species perch on the tops of leaves and periodically circle around their plot, looking for females. They may go to the tops of hills or tall trees, where they aggregate, engaged in what appears to be a flight contest, circling around the meeting point and each other. Such male aggregation behavior (called "lekking") serves to attract unmated females, which makes mating fast and mate choice straightforward.

↑ Qian's Flasher, *Telegonus tsongae*, a tropical skipper found as far north as Texas. It is one of many species of skippers with similar color patterns, which form escape mimicry complexes in the tropics. Skippers also have powerful flight muscles housed in their wide thorax, which allow for fast flight.

SHAPES AND COLORS

Each butterfly uses its wing surfaces for numerous purposes: from thermoregulation to camouflage, from warding off predators to escaping their attention, and from attracting mates to repelling rivals. All of these functions shape the appearance of individual butterflies, so that no two specimens have identical wings, just as no two human hands are exactly the same.

PHENOTYPES VERSUS GENOTYPES

How does the variability of wing patterns correlate with differences in species? How can one distinguish between intraspecific (within species) and interspecific variation? The physical appearance of a specimen is termed its "phenotype" and it can be influenced by both genetic and environmental factors. A specimen's genetic information encoded in DNA comprises its "genotype."

Due to the interplay of genetic and environmental factors, two individuals that look very different are not guaranteed to have commensurate (or even significant) differences in their DNA. So, even with identical genetic starting points, physical outcomes can be quite different, as illustrated by seasonal forms of butterflies (see Chapter 1, page 16).

← A Monarch butterfly, *Danaus plexippus*, with a normal wing pattern versus one affected by a mutation that prevents wing veins from forming. This rare and fatal aberration demonstrates how veins influence wing pattern development.

→ Changes to the underside wing pattern in the Gulf Fritillary, *Dione vanillae*, were experimentally achieved by researchers through CRISPR-Cas9 technology. (CRE means cis-regulatory element; a similar change, shown here, can be achieved by heparin injection shortly after pupation.) See box, page 111, for the attributes bestowed by specific genes.

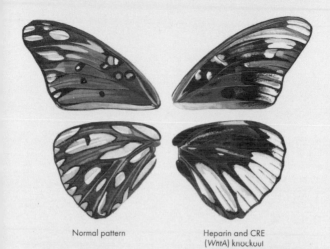

Normal pattern

Heparin and CRE
(*WntA*) knockout

optix knockout

WntA knockout

THE NYMPHALID GROUND PLAN

For all their stunning variety, there are recurring motifs in butterfly wing patterns, which suggest an underlying organizational pattern that applies across species, genera, and even families. Butterfly and moth families (over 120 in the order Lepidoptera) have been traditionally defined in part by similarities in wing vein structure—these variations are small but often enough to distinguish larger taxonomic groups.

Coincidentally, wing veins also play an important role in defining the formation of color patterns. There have been several attempts to propose a "ground plan," where each pattern element is attributed to one or other independently evolving symmetry system.

VARIATIONS ON A THEME

The nymphalid family, with its 6,000-plus species (see Chapter 4, pages 52–57), has provided us with the best models for studying butterfly wing patterns, due to the detailed and regular patterns present on the wing undersides of many species, such as buckeyes (genus *Junonia*) or satyrs (subfamily Satyrinae). These patterns frequently feature concentric eyespots and parallel bands distributed throughout the wing from margin to base. The first ground plan for the nymphalid family was proposed by Boris Schwanwitsch (1889–1957) a century ago, and it has continued to serve as the foundation on which subsequent hypotheses have been built and experiments tested against.

EDIFICATION FROM ABERRATION

It has been known to naturalists for at least a century that subjecting a chrysalis to extreme temperatures can produce wing pattern aberrations. One can get a glimpse into the plasticity and heritability of wing pattern elements through such environmental manipulations, as well as through the classic genetic breeding experiments.

Mutations can shed light on fascinating aspects of physiology and wing pattern development in butterflies. There is a kind of mutation in butterflies that causes them to emerge from their chrysalis without veins in their wings, like sails without a mast. When this occurs in the

SYMMETRY SYSTEMS

The classic nymphalid ground plan is organized into five symmetry systems. The frequently dark-colored areas close to the wing attachment points are called the Wing Root Band System (WRB) and Basal Symmetry System (BaSS). The Marginal Band System (MBS) usually presents itself as a series of thin lines, while the Border Symmetry System (BoSS) frequently features a series of eyespots located between veins.

Toward the middle of the wing, the Central Symmetry System (CSS) overlaps with the Discal Symmetry System or discal spot (DS) where the veins form a closed loop. There, the cross-vein between radial veins allows for hemolymph to circulate through and out of the wing. This area is important for the wing's thermoregulation and circulation, so the wing scales in this region play significant physiological roles, in addition to signaling.

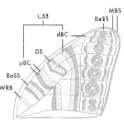

> Nymphalid Ground Plan. WRB (wing root band), DS (discal spot). Symmetry systems: BaSS (basal), CSS (central: pBC and dBC its proximal and distal bands), BoSS (border), and MBS (marginal band).

monarchs and swallowtails, there is a corresponding effect on their color pattern: the individual spots that are normally located between wing veins merge into continuous lines (see page 106). This suggests that spots are not formed independently of one other but are instead tied in some way to the wing vein structure. Yet, in the Squinting Bush Brown, *Bicyclus anynana*, other mutations affect individual eyespots, so there is a complex interplay between genes that determines the layout of symmetry systems and those that control the shape, size, and colors of elements locally.

MANIPULATING COLOR

W e are beginning to understand the general mechanisms of wing pattern regulation through research in the field of evolutionary developmental biology (known as "evo-devo"). In biological organisms, genes are translated into external characteristics via a cascade of events, such as the synthesis of RNA and proteins, and can be turned on and off via epigenetic processes, in which special proteins bind to regulatory regions in DNA and thus control gene expression. The accessibility of these DNA sites is regulated by various processes, such as microRNA transcription, methylation, and histone modification.

WING PATTERN GENES

There are several genes involved in mapping where spots and lines will appear on the wing and what color the background will be. Genetic studies traditionally relied on careful breeding of normal and mutant lineages in the lab and subsequent analysis of phenotypes and genotypes via molecular and mathematical techniques.

CATERPILLARS WITH WINGS

The embryonic wings of a butterfly are small and transparent within a late-stage caterpillar, but already at this stage, the wing pattern is forming. It is mapped out by special genes, such as *WntA* (for stripes), *splat* (for eyespots), and *optix* (for color determination), which are responsible for directing where each set of lines and eyespots will appear and what they will look like. This process reaches completion in the pupal stage when scales develop their ultimate structure and are endowed with their final color.

Today, advances such as CRISPR-Cas9 technology allow researchers to probe what specific genes and their regulatory elements do by "silencing" them (locating and excising small regions of DNA). For example, by silencing regulatory elements in several species of very different nymphalid butterflies, it was recently found that species separated by over 50 million years of evolutionary history, such as longwing butterflies (*Heliconius*) and the Painted Lady, *Vanessa cardui,* have similar genetic mechanisms for controlling wing pattern formation, whereas the Monarch, *Danaus plexippus,* has diverged a lot.

GROUND PLANS FOR OTHER FAMILIES

There is a logical follow-up question: how can the nymphalid ground plan (see page 109), defined by the study of a single family of butterflies, describe the huge diversity of butterfly and moth wing patterns? While it is tempting to come up with theories based on studying "model" species, advances in this field will benefit from studying many species with the help of the laboratory techniques described above, with the nymphalid framework serving as a helpful guidepost. At this time, there are no universally accepted "ground plans" for other families, but it is an active area of research.

← Wing close-up of a Painted Lady. The scale cover gave the Lepidoptera order its name (*lepis* means "scale" in Greek).

PIGMENTS AND STRUCTURAL COLOR

The spectacular colors that help butterflies make an entrance wherever they go are the result of minute scales layered on the wings. Wing scales (see page 110) are chitinous structures covered in ridges and furrows. They each derive from a single cell and are surrounded by support cells in the developing wing.

PIGMENT DIVERSIFICATION

Pigment molecules create colors by absorbing most wavelengths of light and reflecting a small range that we perceive as color. For example, melanins manifest as red, brown, and black colors; pterins, ommochromes, and carotenoids appear yellow or orange. These pigments can undergo rapid evolution to aid survival. For example, the Asian Common Palmfly, *Elymnias hypermestra*, is palatable, but its yellow ommochromes have diversified to mimic toxic milkweed butterflies of the tribe Danaini (see Chapter 4, page 52 and Chapter 6, page 81).

INVISIBLE BUTTERFLIES

There are certain butterflies that completely lack wing scales, reverting to the naked wing architecture found in other insects. Transparent, non-reflective wings can be advantageous in escaping unwanted attention, as demonstrated by the Antillean Clearwing, *Greta diaphanus*, and the Uncolored Clearwing Satyr, *Dulcedo polita*. Investigations on the Glasswing butterfly (now classified as a subspecies, *Greta morgane oto*, see opposite) found that such scale-less wings are covered in irregular, chitinous nanopillars that allow light to dissipate, and researchers have been able to use nanotechnology to recreate artificial structures with a similar function.

↑ Central American Darkened Rusty Clearwing, *Greta morgane oto.*

The non-reflective transparency of this butterfly's wings has been studied in detail.

IRIDESCENT COLORS

Frequently, the apparent color of a butterfly wing can change based on the angle at which it's viewed. Iridescence in butterflies is not due to pigments but rather to nanostructures on wing scales that are on the order of the wavelength of incoming light. Nanostructured wing scales are biologically "cheaper" than producing pigments and can rapidly evolve under selective pressure. For instance, a slight change in scale ultrastructure can shift the wing color from blue to green. And when scientists artificially selected for iridescence in captive butterfly populations, it only took a few generations to produce significantly more iridescent butterflies!

ULTRAVIOLET PATTERNS

Some butterfly wing surfaces reflect UV light (shorter wavelengths below 400 nm), so they may appear very different to other butterflies or birds than they do to us. Humans can only detect UV patterns using special instruments, like spectrometers, or by placing UV lenses on cameras.

FINDING A MATCHING PAIR

To produce fertile progeny, a female butterfly must choose one of its own species, so subtle visual cues, like UV patterns on wings, may have evolved to help the female butterfly invest its time wisely. In two co-occurring species of sulphur butterflies (*Colias*), the ultrastructure of each scale is modified in one of the species, so that UV is reflected from its wings. Genetically, the only difference is a modification in a regulatory DNA region responsible for suppressing the expression of bric à brac (bab), a gene that has been shown in fruit flies to control the thermal plasticity of pigmentation.

↓ UV-reflecting scales on the wings of the Orange Sulphur, *Colias eurytheme* (shot in UV light with a special UV lens) make the butterfly appear brighter than the Clouded Sulphur,

C. philodice, which lacks UV-reflecting scales. To butterflies, UV is just another color that, in this case, helps potential mates differentiate between species more easily.

→ Two *Colias* species, the Orange Sulphur (top) and the Clouded Sulphur, co-occur throughout the United States. Coloration differences between these butterflies, including UV-reflecting scales on the Orange Sulphur, help butterflies make correct mate choices. The genetic regulation behind UV wing patterns has only recently been understood by scientists.

| Sunlight | UV | Sunlight | UV |

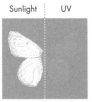

PATTERNS AS DEFENSE

Butterflies of the same species are not the only animals that take cues from wing patterns. Some predators, like birds, rely on a butterfly's appearance when deciding whether to give chase. As a result, one of the main forces that drives the evolution of diversity and bright colors in butterflies is the pressure of predation.

BIRD'S-EYE VIEW

Sporting considerable diversity of their own, birds literally see far and wide. They inhabit every habitat and can hunt butterflies both at rest and in flight. They can also learn quickly and migrate long distances.

To become diurnal, butterflies had to adapt to the challenge presented by birds, and they have done so by developing both cryptic and aposematic (warning) coloration in caterpillars and adults. Studies with frequently uniformly colored *Adelpha* butterflies (collectively known as "sisters") showed that birds can learn to avoid specific coloration, not only of toxic butterflies, but also of faster ones that they have trouble catching.

SEX-LIMITED MIMICRY

In many species, such as the Asian Common Mormon, *Papilio polytes*, and the African Swallowtail, *P. dardanus*, the female can mimic a toxic butterfly, while the male remains non-mimetic and has a completely different wing pattern. How is it possible that two butterflies that share almost all their genetic instructions end up looking unrelated? It appears that sex-determining genes are involved at some point during the formation of wing patterns. Studies point to a single gene (*doublesex*) switching on and off to direct the development of different wing patterns for each sex.

SEASONALLY CRYPTIC

During the wet season, the Common Evening Brown, *Melanitis leda*, which has very prominent eyespots, is reproductively active and is frequently on the wing mating and laying eggs. Experiments have demonstrated that preying mantids selectively target these eyespots, while beak marks are frequently observed on the butterfly's wings in the wild. In contrast, the dry-season generation is much less active and with its nearly absent eyespots, the butterfly relies more on camouflage for safety than misd irection.

↑ The Common Evening Brown, *Melanitis leda*, from South Africa, shown in wet- (top) versus dry-season form. This species is widespread in the Old World tropics and can be a minor pest on rice crops.

WARNING SIGNALS

Bright color patterns are termed "aposematic" when they are meant to be displayed to predators and are strongly associated with an unpleasant taste. Such butterflies fly slowly and rest openly, not counting on a speedy escape but rather on "educating" predators. If a naïve bird tries one of these brightly colored butterflies and gets a bitter mouthful, it will likely avoid its brethren for the rest of its life. Great examples of such patterns are longwings (genus *Heliconius*) and pipevine swallowtails (*Battus philenor*).

BATESIAN VERSUS MÜLLERIAN MIMICRY

Two or more distasteful species may combine their efforts in educating predators by sharing a color pattern, as was first described by Fritz Müller at the end of the 19th century. Sometimes such mimicry complexes consist of species that resemble each other to different degrees. Ithomiine butterflies are defended by alkaloid compounds imbibed with the nectar of certain flowers. Ithomiines and toxic tiger moths (Arctiinae) with similar color patterns are Müllerian co-mimics. In contrast, various similarly colored but edible butterflies, such as metalmarks (Riodinidae) and pierids (Pieridae), are their Batesian mimics (see Chapter 4, page 46).

HUMAN IMPACT

The burgeoning of human civilization across the globe has resulted in profound changes to ecosystems. Together with evolving climatic conditions, the expansion of human communities and the increased use of land has contributed to significant transformations in the landscapes that butterflies inhabit.

SHAPING THE ENVIRONMENT TO OUR NEEDS

For many generations, Native American tribes used fire as a tool to transform their environment: clearing land to attract grazing animals they could hunt and to promote the growth of the desirable plants. As agriculture took hold, first in Anatolia, then in Europe, and nomadic groups established more permanent settlements, they cleared land to grow cereal crops and graze livestock. While alterations to the environment were relatively localized, bones and remains found in early human habitations suggest that the spread of humans often coincided with the extinction of vertebrate fauna. Only recently have we begun to notice the extinction of butterflies and other insects.

AN EXPONENTIAL RATE OF CHANGE

Towns and surrounding farms began coalescing into cities 7,000 years ago, from Mesopotamia to Egypt to Mesoamerica, and the human population began to rise steadily, in spite of war and disease. The Industrial Revolution kickstarted a period of unparalleled technological innovation and improvement in quality of life for mankind. Such a momentous impact on the course of human development was inevitably accompanied by an equally tremendous effect on the biosphere.

Today, in many parts of the world, only small portions of natural terrestrial habitats remain intact, with most fertile lands in use either by towns, forestry, or agriculture. The extensive use of herbicides and pesticides in such settings, as well as the reliance on non-native monocultures of plants, means that there are even fewer places for insects (including butterflies) to flourish.

HABITAT TRANSFORMATIONS

Some habitats formerly housing rich biodiversity are becoming unsuitable for butterfly host plants. The disappearance of glacial mountain tops can turn flowering meadows into xeric rocky outcrops. Reduction in forest cover reduces global rainfall, accelerating desertification.

In protected tropical ecosystems where forest cover has increased in recent years, as in the Guanacaste Conservation Area in Costa Rica, scientists have still observed a marked decline in the number of insects. The root cause of this decline is yet unclear, but decreased rainfall and higher peak temperatures may be responsible. As insects rely on the cyclical appearance of fresh growth on their host plants, any drastic shifts in the timing of rains can affect their reproduction.

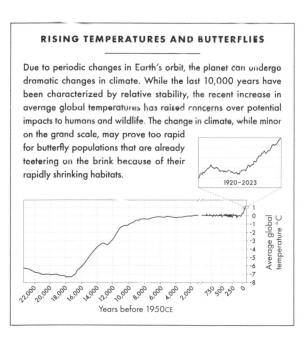

RISING TEMPERATURES AND BUTTERFLIES

Due to periodic changes in Earth's orbit, the planet can undergo dramatic changes in climate. While the last 10,000 years have been characterized by relative stability, the recent increase in average global temperatures has raised concerns over potential impacts to humans and wildlife. The change in climate, while minor on the grand scale, may prove too rapid for butterfly populations that are already teetering on the brink because of their rapidly shrinking habitats.

1920–2023

Average global temperature °C

Years before 1950CE

THE SIXTH MASS EXTINCTION OR . . . ?

The modern, human-dominated age has sometimes been described as the "sixth mass extinction." Indeed, it is impossible to accurately assess how many species of animals and plants have become extinct in the last century, and not least because some have disappeared before they were ever catalogued or identified.

BUTTERFLIES AND HABITAT HEALTH

Entomologists conduct surveys of biodiversity through butterfly and moth trapping in an attempt to monitor the decline of insect abundance and species richness. Butterflies are a great group for such monitoring, because they can be captured, marked, photographed, and released. As a result, butterflies are not only frequently used as flagship species for conservation movements but also serve as a valuable tool for assessing biodiversity and its trends.

GRASSROOTS EFFORTS

Efforts to record butterfly diversity and abundance in Great Britain have been ongoing for over half a century, spearheaded by local butterfly enthusiasts, in contrast to the tropics where, until recently, it was only visiting scientists who were formally noting down observations. Summarized in scientific papers, this information was inaccessible to most and had little influence on conservation locally or nationally. Now, there are many tropical ecosystem studies led by local scientists who are analyzing changes in regional butterfly diversity. In Ecuador's Yasuní National Park, park rangers are surveying their habitat for butterflies in an effort to understand long-term population dynamics. This decentralization of science will be critical for preserving local environments going forward.

WHAT KIND OF FUTURE DO YOU WANT?

Discussions about conservation are often difficult ones as the importance of wildlife and biodiversity depends so much on personal values. Is there an intrinsic value to every butterfly species? What about the aesthetic value or spiritual fulfillment that one gains from being in nature and from watching butterflies? Is there a potential practical value to humanity in the form of yet-to-be-discovered compounds or engineering designs inspired by nature and, if so, how do we weigh this against the value derived from using land for agriculture and resource extraction?

One way to approach such questions is to think about the kind of future we want for ourselves, for humanity, and for Earth. We hope that this book, with its beautiful illustrations and exploration of the complexity and diversity of a single order in the animal kingdom, can contribute to the wider appreciation of butterflies and efforts to conserve them.

BUTTERFLIES AND PESTICIDES

Speaking from personal experience, every time a mosquito-control truck "fogs" for adult mosquitoes in our neighborhood, all the butterflies and caterpillars die too. The effect is immediate and disturbing, as one observes butterflies that were just floating in the air helplessly dying on the ground. Such pesticides tend to work indiscriminately on the nervous system of insects, causing almost instant death, and they can also settle on foliage as a residue, killing caterpillars that feed on it.

THE PRICE OF A MOSQUITO-FREE LIFE

Years ago, one of the authors visited a coastal habitat in the Dominican Republic and then a similar habitat in southern Florida. The difference between the two habitats was striking. The former was teeming with cracker, snout, daggerwing, and purplewing butterflies by day, but in the evening, at an outdoor restaurant, long pants and repellent were needed to ward off the mosquitoes. The reverse was true in Florida: the complete comfort of a mosquito-free evening meal outdoors came at the price of a subtropical paradise stripped of butterflies (and many other organisms).

↓ The *Aedes aegypti* mosquito shown here can transmit dengue and yellow fever. Yet, in the absence of disease, spraying adulticides should be a last resort because it upsets the natural balance of predators and prey in the ecosystem, and greatly reduces biodiversity, including that of butterflies.

→ Butterflies of the Caribbean coastal habitats. From the top: Antillean Snout (*Libytheana terena*), Caribbean Daggerwing (*Marpesia eleuchea*), and Florida Purplewing (*Eunica tatila*). These species used to be common and still persist where mosquito spraying is absent.

CHANGES IN FLORA AND FAUNA

Every ship and plane that crosses the ocean carries with it, not only goods and people, but inadvertently, microbes, seeds, and insects, as well. Since the beginning of global trade, the flora and fauna of Earth has changed dramatically, especially in disturbed habitats and around human settlements, where non-native plants can be prevalent.

Non-native plants can sometimes escape into the wild and become invasive exotics, overrunning entire habitats. Frequently, native butterfly species can adapt to feeding on exotic nectar sources, but for caterpillars, adapting to new host plants is trickier. Exotic plants are either unsuitable for native species, or serve as "biological

INADVERTENTLY
IMPORTED PREDATORS

In the mountains of the Dominican Republic, near the peak of Pico
Duarte at elevations of 10,000 ft (3,000 m), the large and beautiful
endemic Many-spotted King, *Anetia briarea*, gathers in large
roosts for the winter. Scientists have recently noticed intense
predation on these butterflies. While some of these toxic beauties
are nabbed by birds, most birds release them without fatal injuries.
It is likely that the majority are being killed by the invasive black rat
that has finally made it to this part of the island after arriving on
Hispaniola with Christopher Columbus.

sinkholes," where butterflies lay eggs that never fully develop into
adults. This is the case with some species of exotic passion vines and
the Zebra Longwing, *Heliconius charithonia*, in the southeastern
United States. Landscaping with native plants can help sustain
butterfly populations, while providing the concomitant satisfaction
of butterfly-watching.

PSEUDO-FORESTS

Throughout the world, from Brazil to New Guinea, one can
observe dense "forests" that entice a naturalist from a distance. Upon
closer inspection, however, one realizes that these are exotic
eucalyptus, oil palms, or other non-native monocultures. Very few
butterflies will be found here (although if there is some grass growing
in these plantations, one can sometimes find
common satyrnine and skipper species, as a
consolation). Similar situations may occur in
more temperate habitats where hardwoods
have been replaced with conifers.

← The Zebra Longwing
is one of the most
common butterflies
in suburban habitats
of the southeastern
United States due to
the popularity of its
host plant, the passion
vine (*Passiflora*), with
gardeners.

BUTTERFLIES IN MONOCULTURES

There are several butterfly species that can survive and occasionally even thrive in monoculture environments (although moths are more infamous as pests of commercial crops).

WHO ATE MY . . .?

The Common Evening Brown, *Melanitis leda*, can reproduce in a rice field, and the cabbage-feeding whites can thrive among cruciferous crops. The Giant and Lime Swallowtails (*Heraclides cresphontes* and *Papilio demoleus*) use cultivated citrus as host plants, and the caterpillars of the North American Zebra Swallowtail, *Eurytides marcellus*, strip the leaves of paw-paw fruit trees.

Bananas in Latin America may promote the populations of several owl butterfly species of the tribe Brassolini. On the island of Hispaniola, caterpillars of *Calisto pulchella*, a small brown satyrine with orange patches, make a niche for themselves by scraping away the surface of the sugarcane stem concealed by the sheaths of the surrounding leaves. In your backyard, you might occasionally find your parsley, carrots, or dill crawling with the caterpillars of the Black and Old World Swallowtails, *Papilio polyxenes* and *P. machaon*.

↓ *Calisto pulchella* caterpillar feeding on sugarcane, introduced to Hispaniola at the end of the 15th century.

↓ The caterpillars and adults of owl butterflies (*Caligo*) love bananas, which were introduced to South America in the 16th century.

→ The Lime Swallowtail, *Papilio demoleus*, which is originally from Australia/New Guinea, has spread throughout the tropics, reaching Florida in 2022. It feeds on citrus, which was first cultivated by humans 3,500–5,000 years ago in Australasia and introduced to the New World at the end of the 15th century.

HOW BUTTERFLIES ADAPT

I nsects, including butterflies, tend to adapt to change quickly. They can survive in small patches of remaining habitat for a long time and then repopulate available habitats where extinction of other populations has occurred.

DISTURBED VERSUS UNDISTURBED HABITATS

In the 1960s, a multi-year survey of the highly disturbed locality near Lagos, Nigeria, comprised of fields, roadsides, secondary growth forest, and cacao plantations, documented over 300 butterfly species—half of the number found in the country overall. A study in the Amazon showed that forest edges and secondary forests with intermediate disturbance have the highest butterfly diversity.

While we may think of primary habitats, such as virgin forests, as the true natural paradise, even in such habitats, it is the fallen trees that allow for maximal variety by bringing sunlight to the forest floor and room and resources for new plants to thrive. The fresh growth on these plants and their diversity attracts butterflies. When forests are fragmented or selective logging occurs, there is a similar effect: more types of niches are created for different species to inhabit.

BUTTERFLY PHENOLOGY

As average global temperatures rise, the distribution of many butterfly species will change, together with their phenology. In the Northern Hemisphere, this means that species of butterflies found farther south will be flying earlier in the season and farther north. A study of over 100 species of butterflies in Korea has demonstrated northward shifts for many species in the last 50 years. While some species will adapt by expanding their range, scientists warn that many species will simply decline due to the lack of suitable habitats for them to expand to when things heat up.

SHIFTING HOST PLANTS

Butterflies sometimes benefit from human expansion by shifting their host plants to species that we like to grow for food or as ornamentals (see page 126). In and around New Guinea, the gorgeous Palmfly butterfly, *Elymnias agondas*, has switched from native palms to cultivated ones, such as oil palms, greatly expanding its range and populations. The Cassius Blue, *Leptotes cassius*, feeds on plumbago and the Juniper Hairstreak, *Callophrys grynneus*, thrives on ornamental juniper trees, providing good examples of where our cultural practices can contribute to butterfly well-being. The introduction of the European *Plantago* plants to North America increased the numbers of the Common Buckeye, *Junonia coenia*, found along roads where this weedy plant became very common.

→ Cassius Blue lays its eggs on white plumbago, which is native to coastal habitats along the Gulf of Mexico, but also feeds on blue cultivars that originated in Africa.

PEACEFUL COEXISTENCE

Studies of biodiversity in Japan and New Guinea have demonstrated that butterflies and traditional, small-scale family agriculture are quite compatible. For those interested in butterfly conservation, this should give hope and serve as a call to action. In addition to creating national, state, and local parks that serve as reservoirs for native wildlife, especially in special habitats where highly endemic species thrive, each of us can focus on adjusting our habitual practices of land use. Small changes in our local communities can have a big impact on butterflies: planting pollinator-friendly plants between crops, letting "weeds" grow where possible, and minimizing the use of pesticides and herbicides are a few examples.

MYTH AND FOLKLORE

The dramatic transformation from earthbound caterpillar to airy, lithe butterfly, along with their unique migratory behaviors and their eye-catching wing patterns, gives rise to stories and myths about the provenance and meaning of these otherworldly creatures.

SPIRITS OF THE DEAD

In Mexico, the Day of the Dead celebrations take place in the beginning of November, just as millions of monarch butterflies (*Danaus plexippus*) are finally reaching the endpoint of their long migration route. To many in these regions, like the Purépecha, an indigenous group from Michoacán, and the Mazahuas from the state of México, these butterflies represent the souls of deceased ancestors that are being welcomed home over these two days.

PSYCHE AND SOUL

In the Greek language, the word "psyche" means both "soul" and "butterfly," an association reinforced by the myth of Eros and Psyche, in which Aphrodite's son falls for a girl who has incited the jealousy of the goddess. Jacques-Louis David depicts the lovers in his 1817 painting *Cupid and Psyche*, with a cabbage white butterfly hovering suggestively over the human Psyche.

In Chad and Senegal, the entrance of butterflies and moths into a home can signify a visit from an ancestor, compelling the belief that these creatures should not be harmed. In the Sahel region of Africa, the names of butterflies are often religious, referring either to God or religious leaders. Swarms of butterflies are interpreted by some ethnic groups, like the Bwa of Burkina Faso, as signs of divine approbation, hailing the start of the rainy season.

BUTTERFLY WARRIORS

In Mesoamerican cultures, butterflies have an unexpectedly martial connotation and were symbols of fire, warriors, and rebirth. Butterfly symbols can be found on the murals and pottery of Teotihuacán (see page 132). In Tula, the capital of the Toltec people from 850–1150, there are four towering Atlantean warrior figures with stylized butterflies depicted on their breastplates. The Aztecs, a notoriously martial people, believed butterflies to be the souls of warriors slain in battle.

Among the Rukai people in Taiwan, the title of *lyalivarane*, which means "butterfly," is granted to those who are the fastest, earning them the right to wear the butterfly headdress. In Paiwan tradition, nimbleness is rewarded with swallowtail beads, and prowess at weaving qualifies women to wear dresses with butterfly patterns.

← The Day of the Dead Parade, Mexico City. Dancers celebrate the mass migration of Monarch butterflies to the Sierra Madre Oriental from the eastern USA and Canada.

BUTTERFLIES IN ART

D ue to their ubiquity across the world, butterflies appear in the artwork of many cultures and ages. In the visual arts, they can symbolize the soul, resurrection, transformation, and fragility, allowing artists to explore themes such as man's relation to nature, human potential, and the ephemerality of life.

NATIVE AMERICAN ART

Butterflies are frequently depicted in the petroglyphs, ceramics, and jewelry of the Native American people of the North American Southwest. Naturalistic insect designs, including butterflies, feature in the black-and-white Mimbres pottery produced by the peoples who once inhabited the Mimbres River Valley in New Mexico and Arizona. Dry paintings, also known as "sandpaintings," are important components of Navajo ceremonial rituals, and many insects, such as butterflies, are included for either their symbolic or mythological significance. These artworks are ephemeral and created with sand, crushed rocks, and dry plants of different colors arranged in patterns on the ground.

RENAISSANCE AND BAROQUE ART

In the Renaissance and Baroque periods, butterflies often featured as depictions of the human soul, while their metamorphosis had religious connotations of resurrection. Not all imagery was meant to be uplifting, however: in Hieronymus Bosch's work, *The Garden of Earthly Delights* (ca. 1503), it is not angels but devils who sport butterfly wings—one has the pattern of a Small Tortoiseshell, *Aglais urticae*, and the other, the eyespots of the Meadow Brown, *Maniola jurtina*.

↓ Butterfly imagery from Teotihuacán (see page 131). From souls to rebirth, butterflies had many meanings in pre-Columbian America.

Butterflies and caterpillars were popular additions to still-life paintings, such as *Still Life with Flowers in a Vase* (1617) by Dutch painter Christoffel van den Berghe, which includes several

common European butterfly species. Apart from showing off an artist's skill in illustrating delicate insect anatomy, they imparted symbolism to seemingly mundane objects.

JAPANESE WOODBLOCK PRINTS

Katsushika Hokusai, one of the best-known ukiyo-e painters of the Edo period (1603–1867), who produced universal classics like *Under the Wave off Kanagawa*, featured butterflies in his woodblock prints. An especially evocative one is *A Philosopher Watching a Pair of Butterflies* (1809–1819), which invites the viewer to ponder the insects alongside this scholar-warrior. Hokusai's *Peonies and Butterfly* (ca. 1833) is strikingly dynamic despite its precise lines: the strong wind is felt through the bowed wings of the butterfly that is valiantly facing the gust, perhaps a symbol of persistence in the face of daunting challenges.

Kamisaka Sekka, a master of the Rinpa style of painting, which favors motifs of nature and the seasons, published two volumes of woodblock prints devoted to butterflies. *Chō senshu*, or *A Thousand Butterflies* (1904), is a series of minimalist woodblock prints in which butterflies of all shapes, colors, and sizes (some real, most imagined) are the stars. The kaleidoscopic variety of butterfly wing patterns in nature likely inspired Sekka to imagine the myriad possibilities of these elements in the decorative arts.

BUTTERFLIES IN LITERATURE

Vladimir Nabokov, a luminary of 20th-century literature and a lepidopterist, wrote: "Literature and butterflies are the two sweetest passions known to man." In his own works, including *Pale Fire*, butterflies frequently feature as motifs. Their latent symbolism is irresistible to the creative mind, and so butterflies and those who love them find their way into the arenas of art, literature, and philosophy.

THE PHILOSOPHICAL BUTTERFLY

"Zhuang Zhou Dreams of Being a Butterfly" (ca. 300 BC) is a famous story from Daoist philosophy, in which the protagonist, upon waking, cannot be sure whether he is a human who dreamed he was a butterfly or a butterfly who is dreaming of being human. Kafka's *Metamorphosis* (1915) makes one shrink from the thought of being turned into an insect, but this short tale is a more palatable thought experiment of the limitations of one's perception and the impermanence of all earthly distinctions.

THE LYRICAL BUTTERFLY

Just as they are to painters, butterflies are a frequent inspiration to poets. The following haiku by Japanese poet Matsuo Bashō (1644–1694) may be short, but it flies true as an arrow to the heart:

A caterpillar,
this deep in fall—
still not a butterfly.

Emily Dickinson (1830–1886) features butterflies in a number of her poems, such as "The butterfly obtains" and "From Cocoon forth a Butterfly," seemingly drawn to their contradictory themes of beauty versus vanity and apparent aimlessness versus latent potential.

THE PROVERBIAL BUTTERFLY

Butterflies feature in the proverbs of various ethnic groups across the African continent. For the Fon people of Benin, the proverb *Awadakpɛkpɛ wɛ un nyi* (meaning "I am a butterfly; I fly from flower to flower") expresses the freedom and liberty associated with butterflies. The Yoruba people of West Africa, no doubt observing the erratic flight of some species, have a saying: *Yio b'ale, yio b'ale ni labalaba fi wo'gbo* (which means "It will settle down, it will settle down, yet still the butterfly flies into a bush thicket"). This spasmodic behavior may be why in Rwanda and Chad calling someone a butterfly connotes having bad manners or being untrustworthy.

← Ike no Taiga (1723–1776), of Kyoto, Japan, was among many artists to illustrate the "Dream of a Butterfly" episode from the *Zhuangzi* (ca. 300 BC), an ancient Chinese text foundational to Daoist philosophy. This tale is associated with the themes of transformation and the illusory nature of reality.

COMMERCIAL USES

Butterfly souvenirs—mounts of colorful butterflies behind glass, jewelry made from iridescent wing fragments, and collages composed of butterfly wings—can often be found in travels to the tropics. These crafts help local people capitalize on their natural resources and embrace ecotourism.

LIVE BUTTERFLY EXHIBITS

In the 1970s, the first live butterfly exhibits appeared in England, Malaysia, and Florida. These enclosures are a combination of a garden and a zoo, where one can stroll among tropical plants and watch butterflies feeding on flowers and fermented fruit. Behind this simple concept is a complex operation of butterfly farms in the tropics that harvest chrysalides and ship them to exhibit facilities in time for the butterflies' emergence. These fragile businesses provide an important incentive for people in highly biodiverse regions of tropical countries, such as Costa Rica, Ecuador, and Malaysia, to conserve the local butterfly habitat as a source of stock and host plants.

↓ Neotropical Blue Morpho, *Morpho peleides* (wingspan 5–8 in / 13–20 cm), and Australasian Cairns Birdwing, *Ornithoptera euphorion* (wingspan 7–10 in / 18–25 cm). Large, colorful species are popular with breeders supplying live butterfly displays.

→ Prola Beauty, *Panacea prola* (wingspan 2½–3 in / 6–8 cm), with its striking iridescent blue wing pattern and blood-red underside, is a spectacular addition to many butterfly houses. Here it is shown perching on a leaf in the Butterfly Rainforest exhibit at the Florida Museum of Natural History.

ORIGINS OF LEPIDOPTEROLOGY

Most of what we know about butterflies today, from their classification to their ecology and distribution, comes from people who collected specimens. Conservation efforts are often pioneered by people who engage in butterfly collecting, just as fishers and hunters are often the strongest proponents of conserving nature.

CABINETS OF CURIOSITIES

In the 16th century, as humanist philosophy and Baconian ideas about the virtue of amassing knowledge of the world spread among European nobility, it became fashionable to collect objects for display in *Wunderkammer*, also known as "cabinets of curiosity." An ideal collection would contain a balance of items drawn from nature, unusual natural phenomena, objects manufactured by humans, cultural paraphernalia from all corners of the world, scientific instruments, and historical artifacts. Butterflies became popular additions to these cabinets of curiosity due to their beauty, associations with exotic lands, and ease of display in pinned form.

RECORDING NATURE

↓ Assembling butterfly collections in beautiful wooden cabinets was fashionable among the wealthy elite in the Victorian age.

While employed as court artist to the Holy Roman Emperor Rudolf II, who himself had one of the most extravagant cabinets of curiosities, Joris Hoefnagel (1542–1601) produced stunningly naturalistic depictions of butterflies and caterpillars in natural poses as illuminations to manuscripts. His four-volume collection, *Four Elements*, contained 300 miniatures meticulously executed on parchment with watercolors. Taxonomic nomenclature was only developed two centuries later, but such detailed and accurate illustrations of the natural world likely contributed to advances in the field.

INSECTS OF TROPICAL PARADISE

A daughter of a Dutch engraver and trained by her stepfather (himself a student of the German still-life painter, George Flegel), Maria Sibylla Merian (1647–1717) further ignited the European public's imagination about tropical insects. She added to the store of entomological knowledge with her drawings recording the exquisite beauty of the tropical insect fauna she observed in her travels to Suriname. One of her favorite themes was insect development, to which she devoted her *Metamorphosis Insectorum Surinamensium*, where she featured not only adults but also the immature stages of butterflies.

THE FIRST CITIZEN SCIENTISTS

Unlike many other groups of insects (with the possible exception of larger beetles), the study of Lepidoptera, including the identification of new species, descriptions of life histories, and cataloguing of distribution, has largely been spearheaded by amateur lepidopterists. Most of the vast collection holdings in major museums come from the donations of specimens collected by such amateurs over their lifetime. Those interested in Lepidoptera, both professionals and amateurs, are frequently united in Lepidopterists' Societies, organizations that exist on both the regional and international scale.

FAMOUS LEPIDOPTERISTS

Vladimir Nabokov (1899–1977), an author referenced earlier in this chapter, was a serious taxonomist who specialized in the lycaenid tribe Polyommatini, and to this day many of his specimens can be found at the Museum of Comparative Zoology at Harvard University. Women often had a more difficult time being accepted into the world of natural history exploration, but this did not stymie Victorian lepidopterist Margaret Fountaine (1862–1940), who collected butterflies in over 60 countries, often undertaking the work alone. She was elected a member of the Royal Entomological Society and, on her death, bestowed a stunning collection of specimens and scientifically accurate watercolors.

BUTTERFLY GARDENING

Victorian explorers brought many of the interesting plants that they encountered in their travels back to England, and these exotic blooms started making their way into gardens. Among them was the butterfly bush (*Buddleja davidii*), which was imported from China in 1774, and has today become a staple of "butterfly gardening"—the practice of selecting plants with the primary purpose of attracting Lepidoptera. In recent years, butterfly gardening has taken off as a phenomenon in Europe and the United States.

HOST PLANTS AND NECTAR SOURCES

Butterfly host plants provide the opportunity for butterflies to lay eggs and for caterpillars to develop. They tend to be unique to each species, so should be tailored to butterflies found in your area. Nectar, on the other hand, can come from a wide variety of plants. Composite flowers such as daisies, black-eyed Susans, and coneflowers, provide a convenient landing platform for larger butterflies. Many flowers with deep corollas, such as the firebush, appeal to butterflies with long proboscises.

↓ A satyrine, *Melanargia galathea*, visiting a butterfly garden outside of Grenoble, France.

↓ Fritillary perched on *Rudbeckia*, a common garden flower that is also an excellent nectar source for butterflies.

→ Attracting the Scarce Swallowtail, *Iphiclides podalirius*, to a European butterfly garden can be achieved by planting a combination of butterfly bush (*Buddleja*) and blackthorn (*Prunus spinosa*), with the former providing a nectar source for the adults and the latter food for its caterpillars.

THE BUTTERFLY OLYMPICS

Human athletes compete for glory, butterflies—for survival. As a result, butterfly species vary greatly in their abilities, and this impressive range, which merely reflects the variety of survival strategies, can be appreciated best by considering the extremes.

DISTANCE FLYING

We have already discussed the fact that some butterflies can traverse immense distances when highlighting the migration of monarchs to Mexico (see Chapter 8, page 102). The longest migration record, however, belongs to the Painted Lady, *Vanessa cardui*, which migrates from the Arctic to its breeding grounds in Africa, covering over 10,000 miles (16,000 km). On the other extreme are homebodies like the female Anicia Checkerspot, *Euphydryas anicia*, of Colorado which only leaves its host plant patch where it was born when harassment by males becomes unbearable.

↓ Painted Lady butterflies in Colorado, feeding on a flowering rabbitbrush, one of over 300 nectar plants that they reportedly use while migrating across North America.

GROWTH RATE

Among the fastest to reproduce are the carnivorous lycaenid butterflies. Meat-eating speeds up their development considerably, as is the case for the caterpillars of the Harvester butterfly, *Feniseca tarquinius* (see Chapter 5, page 62), which can become fully grown in a week on their diet of aphids. In contrast, some butterflies develop extremely slowly, especially if they feed on grasses. For instance, the Squinting Bush Brown, *Bicyclus anynana*, develops from egg to adult in about 40 days when temperatures are optimal (at 81°F/27°C), but decreasing the temperature by just 7 degrees almost doubles that time.

SPEED

Clearwing butterflies (Ithomiinae; see Chapter 9, pages 112–113) are probably among the slowest-flying butterflies, spending most of their time hovering in place. They have short, slender thoraxes and their wing movement is frequent but shallow in amplitude. In the same Neotropical habitat, you may see other nymphalids, such as iridescent *Prepona*, with triangular, short, and broad forewings and very thick, long thoraxes in proportion to their wing length. These thoraxes house powerful wing muscles that enable the butterflies to fly at speeds exceeding 30 mph (48 km/h).

SIZE

The 6-in (15-cm) wingspan of the Jamaican Homerus Swallowtail, *Papilio homerus*, the largest butterfly in the New World, pales in comparison with that of the Queen Alexandra's Birdwing, *Ornithoptera alexandrae*, from Papua New Guinea, which can be double the size. The smallest butterflies are in the Lycaenidae family. The wingspan of the Western Pygmy Blue, *Brephidium exilis*, which is naturally occurring in the western United States and the Caribbean and has recently been introduced to the Old World, does not even exceed half an inch.

BUTTERFLY DRUNKS AND COMMUNAL SLEEPERS

On a collecting trip, the Neotropical Cassia's Owl Butterfly, *Opsiphanes cassiae*, was observed to fly into a room and sip on the wine that had been spilled on the table. This large butterfly got inebriated very quickly and became disoriented, flying in a spiral pattern and flapping its wings ineffectually upon landing. In a few hours, it had slept it off and flown away, acting as if nothing happened. These butterflies are naturally attracted to fermented fruit, but they don't normally imbibe enough alcohol to become inebriated.

DRINK TOGETHER, SLEEP TOGETHER

Many butterflies end up drinking together from wet ground and fermented fruit. They are drawn to the smells of alcohol and ammonia but also take their cues from each other and segregate by species when puddling. Some butterflies take social obligations to another level: at night, in the darkness of the rainforest, daggerwings (*Marpesia*) and longwings (*Heliconius*) sleep together, hanging off branches like grapes from dusk until dawn.

↓ Fruit fermentation by-products are attractive to fructivorous butterflies, like this Cassia's Owl Butterfly.

↓ The Berania Daggerwing, *Marpesia berania*, roosts in groups overnight in the rainforest, gaining communal defense.

→ *Heliconius erato phyllis* in Argentina form overnight roosts, which is common in this genus. New roost members are "recruited" at flowers by older butterflies, whom they follow to the roost sites. The exchange of foraging information can be an additional reason for this gregarious behavior.

EXCREMENT FEEDING AND FLINGING

Both juvenile and adult skippers (Hesperiidae, see Chapter 3, pages 42–45) have an interesting relationship with excrement, but in different ways. Adults like to mix up their more traditional flower diet by feeding on bird droppings, while caterpillars produce their own cannon fire by discharging their excrement pellets (known to entomologists as "frass") with prodigious force.

FRASS PROJECTILES AIDING IN SURVIVAL

The Silver-spotted Skipper, *Epargyreus clarus*, hones its frass-flinging talents as the caterpillar grows bigger and stronger, until it can achieve distances of over 32 in (80 cm) with its compact, 1½-in (4-cm) long body. While one may suppose that this behavior evolved simply as a matter of personal hygiene, it has been hypothesized that it also deceives predators and parasitoids. In a study, caterpillars that had frass in and around their shelters were more easily detected by foraging predatory wasps than those without frass. This is probably even more true for parasitoid wasps (see Chapter 6, page 78), many of which are highly specialized in finding and laying eggs inside live caterpillars.

FLINGING THE FRASS

The caterpillar of the Brazilian Skipper, *Calpodes ethlius*, has translucent skin, which makes it an especially suitable subject for studying the mechanism by which skipper caterpillars expel their frass in such a unique manner. The caterpillar builds up hemolymph pressure by squeezing its posterior prolegs, but rather than transferring the pressure gradually to the excretory system, it releases it all at once. There is a special plate, known as an anal comb, situated just above the anus, which allows the pressure to be contained until jettison. Once that latch is released, the frass pellets, which weigh about 10 mg, are discharged at a velocity of 4.27 ft (1.3 m) per second for a distance of over 28 in (70 cm).

THE "TRICKLE-DOWN ECONOMICS" OF BIRD FECES

Bird droppings may be a nuisance when they land on your prized jacket as you head to the theater, but they are a much sought-after source of nutrients for other animals. The white splotches derive their color from the uric acid excreted by birds and are also rich in salts, phosphorous, and nitrogen—valuable and rare commodities in the natural world. Skipper males make a beeline for them to bolster their reproductive chances, not only to keep their female-searching flight in fine form, but also to pass these commodities on to females through spermatophores during mating.

This fondness for bird feces leads to an interesting cascade effect in the tropics. Birds frequently follow the movement of army ants through the forest, since they can reliably forage on the insects that scatter out of the path of the marching hoard. Skippers also follow the army ants, but they couldn't care less about them—they're just in it for the precious bird droppings that happen to follow the same trajectory.

← Skipper butterflies have a long proboscis and flexible abdomen, which helps them "recycle" water while diluting and imbibing bird droppings.

THE BUTTERFLY MATERNITY WARD

Most butterflies lay eggs either singly or in small batches on their host plants. Many butterflies, from hackberry butterflies (*Asterocampa*) in the United States to many ithomiines in the Neotropics to *Eumaeus* lycaenids, oviposit gregariously, meaning that they lay eggs in batches. By doing so, they give their caterpillars a healthy start that allows them to break through the plant's mechanical defenses: they work together to clear away the hairs or thick epidermal layer of the leaf and get to the good, juicy part.

EVEN MORE EGGS IN ONE BASKET

Some butterflies take caterpillar nurseries to a different level. Remarkably, at least two nymphalid butterfly genera, *Heliconius* and *Aglais*, have evolved a behavior wherein females cooperate in ovipositing together on the same leaf, creating a kindergarten for their young. In tortoiseshell butterflies (*Aglais*), this leads to enormous communal nests of several hundred caterpillars all feeding in the same nettle patch.

↓ Large piles of eggs have advantages, such as outer eggs providing a protective shield against parasitoids.

↓ Gregarious oviposition leads to large groups of caterpillars feeding together, overcoming plant defenses.

→ Female Indian tortoiseshell butterflies (*Aglais caschmirensis*) sometimes oviposit in tandem on nettles, forming even larger egg piles. Their caterpillars can form large silk nests in nettle patches, which help them stay protected and warm. Older caterpillars venture out to feed and pupate alone.

FOOLING THE SPIDERS

Butterfly coloration is frequently attributed to predation by birds. And indeed, there are many instances of aposematic coloration to warn of toxicity, as well as mimics of these patterns, which affect avian predatory behavior and thus butterfly evolution.

There are also many documented cases of false-head patterns (see Chapter 5, pages 64–65)—where a tail and a fake eye on the hindwing simulate the head—that may cause birds to attack the "wrong end" of the butterfly, allowing it to escape. This clearly works well for some larger butterflies, such as the Scarce Swallowtail, *Iphiclides podalirius* (see Chapter 11, page 141), which frequently loses its tails to birds. However, many of these false-head patterns are found in very small butterflies (and moths) and seem too minute to effectively misdirect bird attacks. So what purpose do these adornments serve?

JUMPING SPIDER VISION

The reason the false-head distraction works so well on the jumping spider is most likely due to the unique nature of spider vision. Jumping spiders are especially adept at both front-focused color vision (provided by the two principal eyes) and 360-degree, gray-scale, peripheral vision that alerts them to any signs of movement. When met with a hairstreak, the spider focuses on the butterfly like a normal target, but is likely distracted by the bright-colored spots on the "tail" moving up and down. This shifts the focus of the principal eyes (and hence of their attack) from the vulnerable head region to the "disposable" wing region. In nature, one can often find hairstreak butterflies with missing false-head regions—no doubt survivors of spider attacks.

HAIRSTREAKS VERSUS SPIDERS

In a series of showdowns orchestrated in the laboratory, a *Phidippus* jumping spider was paired with various moths and butterflies. Each time the spider was presented with its prey, it knew exactly where to pounce to deliver a paralyzing bite directly behind the head. It would then triumphantly suck its victim dry. However, when this undefeated contender was presented with a Red-banded Hairstreak, *Calicopis cecrops*, which has a false-head pattern, the spider aimed for the false-head region every time. After a while, the spider just gave up on the hairstreaks, showing how effective this strategy is against small-scale predators.

↑ Many species of Lycaenidae, such as this Red-banded Hairstreak, defend themselves against jumping spiders by deflecting their attacks from their real head to the "false head" on their wings.

GENDER, SEX, AND CHASTITY BELTS

There are many unusual rituals in the mating of butterflies, such as the advent of chastity belts, pupal mating, and skewed sex ratios. Though rare, intersexes and gynandromorphs can really stand out from the rest with their part-male, part-female wings.

SPHRAGIS: THE ULTIMATE CHASTITY BELT

In close to 300 species of butterflies, mating plugs or sphragides are placed by males on or inside the female abdomen during mating. This type of "possessiveness" often manifests in species with less elaborate courtship behavior—the last to mate usually becomes the father, so males try to prevent the female from any future pairings.

Butterflies of the genus *Parnassius* set a record in this regard, with a staggering 20 percent of the male's bodyweight "invested" in sphragis production, which occurs in the accessory glands of male genitalia and constitutes a large investment of proteins and lipids. The waxy, viscous substance is produced gradually during mating and then hardens upon contact with the air, sometimes forming very large and cumbersome-looking structures. Such sphragides are visible to other males, which may prevent further harassment (and certainly prevent physical mating), allowing the female to go about her business of feeding and egg-laying.

PUPAL MATING

Eighteen species in the genus *Heliconius* belong to the "pupal mating clade," in which males seek out females while they are still inside their chrysalides. In the case of the Zebra Longwing, *Heliconius charithonia*, these females don't really get a chance to spread their wings before becoming pregnant. As the female matures inside the chrysalis, the males, which are already on the wing, take note of the pupa and start paying it visits, attracted by both visual cues and smell. Two males can land on opposite sides of the pupa and prevent competing suitors from landing by opening their wings when they approach.

This strange ritual can go on for days until the female is ready to emerge. But before she can do so, the alpha male punctures the chrysalis with its abdomen and mates while the female is still inside. Still joined with the male, the female manages to break out of the pupal shell, but she remains attached to him while spreading and drying her wings. In addition to the spermatophore, the male gives the female anti-aphrodisiac compounds that render his bride unattractive to other males, so that once they separate, she goes about laying eggs without interference from the opposite sex.

The Clouded Apollo, *Parnassius mnemosyne*, is blocked from subsequent mating by a sphargis.

GYNANDROMORPHS AND INTERSEXES

In most animals, if two spermatozoids carrying opposing sex genes fertilize a rare egg cell with two nuclei, it leads to early embryo death. But not so in butterflies. Once in a while, perhaps less than one in 100,000, a spectacular bilateral gynandromorph—a half-male, half-female individual—results from such fertilization.

Intersexes are more common than gynandromorphs, and typically manifest as male or female individuals with discrete wing and body regions that look like those of the opposite sex. Three wings, for example, will be mostly male and one mostly female, so these types of butterflies can also be quite noticeable and dramatic in sexually dimorphic species. Biologically, the entire organism, in such cases, may be either male or female, showing only secondary sexual characteristics of the opposing sex. The cause of the intersex phenomenon can range from mutations to hybridization between different species to infections.

GLOSSARY

abdomen
The last 10 segments of a butterfly's body; in the caterpillar, this is the region from which prolegs protrude.

antennae
Multisegmented head appendages with sensilla responsible for smell in caterpillars and adults; in monarchs, these organs are also used for orientation.

bursa copulatrix
A sac inside the female's abdomen into which a spermatophore is delivered during mating.

chitin
Hard, flexible, polysaccharide-based material from which butterflies make their exoskeleton.

chrysalis
(*pl. chrysalides*)
The pupal stage of a butterfly.

corpus allatum
(*pl. corpora allata*)
A gland that makes juvenile hormone.

cremaster
The tip of the abdomen within a chrysalis, which is equipped with hooks that are used to attach it to a silk pad laid down beforehand by the caterpillar.

CRISPR-Cas9
Clustered Regularly Interspaced Short Palindromic Repeats are used to recognize (and Cas9 enzyme, to cut) target DNA sequences.

epipharynx
Mouthpart where the caterpillar tastes food; in adult butterflies, it is located behind the proboscis.

epithelium
Cell layer that lines the digestive tract in caterpillars and butterflies.

eyespot
A bull's-eye-shaped pattern element on the wing with a distinct center, sometimes with concentric circles around it.

instar
A stage of caterpillar between moltings.

larva
Growing and molting stage of development; synonymous with "caterpillar" in butterflies and moths.

Malpighian tubules
Excretory organs in the abdominal body cavity.

metamorphosis
Transformation from an egg to an adult via caterpillar and pupal stages.

ommatidium
(*pl. ommatidia*)
A unit of compound eyes with its own lens (cornea).

opsins
Pigments that line the bottom of the ommatidium and are responsible for capturing light, making them critical to color vision in butterflies.

osmeterium
(*pl. osmeteria*)
A gland that serves a defensive function and is found behind the head in swallowtail butterfly caterpillars.

ovipositor
The external part of the female's genitalia through which the egg is delivered.

palps
(*pl. also palpi*)
Mouthparts represented by segmented appendages.

proboscis
A coiled, straw-like mouthpart (formed by two halves called galea) through which butterflies drink.

pupa
A relatively inactive stage of development between the caterpillar and adult stages; synonymous with "chrysalis" for butterflies.

scale
A chitinous microscopic structure that serves as a single "shingle," a multitude of which cover butterfly wings.

sclerite
Individual chitinous structures in the exoskeleton that, like our bones, provide support and attachment points for muscles.

sensillum
(*pl. sensilla*)
A sensory organ in insects, including butterflies.

seta
(*pl. setae*)
A stiff, hairlike structure.

spermatheca
A sac in the female's abdomen in which sperm is stored prior to fertilizing the eggs.

spermatophore
A mass that includes sperm and nutrients that is transferred from male to female during mating.

sphragis
(*pl. sphragides*)
In butterflies, this structure is secreted by the male during mating to prevent future mating by the female.

spinneret
An organ located below the caterpillar's mouthparts, where the silk gland opens and silk is spun.

spiracle
The external opening of a trachea.

thorax
Part of an adult butterfly where the wings and legs are attached; the first three (fused) segments of the body.

trachea/tracheoles
Hollow branching tubes that form the respiratory system.

Vogel organ
The hearing organ of nymphalid butterflies located on the wings.

FURTHER READING

Chapter 1 Darwin, C. 1859. *The origin of species by means of natural selection ...*

De Jong, R. 2016. Reconstructing a 55-million-year-old butterfly ... *European J. of Entomology.*

Dobzhansky, T. 1970. *Genetics of the Evolutionary Process.*

Grimaldi, D., and Engel, M. S. 2005. *Evolution of the Insects.* Cambridge Un. Pr.

Kawahara, A.Y. et al. 2023. A global phylogeny of butterflies ... *Nature Ecology & Evolution.*

Lukhtanov, V. A., et al. 2016. DNA barcodes as a tool in biodiversity research... *Systematics and Biodiversity.*

Mallet, J. 2001. Species, concepts of. *Encyclopedia of biodiversity.*

Mayr, E. 1942. *Systematics and the Origin of Species ...*

Chapter 2 Cox, C. B., et al. 2016. *Biogeography ...* Wiley & Sons.

Rosser, N., et al. 2021. The Amazon River is a suture zone ... *Ecography.*

Sourakov, A., and Chadd, R. W. 2022. *The Lives of Moths ...* Princeton Un. Pr.

Sourakov, A., and Zakharov, E. V. 2011. "Darwin's butterflies"? DNA barcoding and the radiation of ... genus *Calisto* ... *Comparative Cytogenetics.*

Chapters 3–5 Ackery, P. R., and Vane-Wright, R.I. 1984. *Milkweed butterflies ...* British Museum.

Carvalho, A. P. S., et al. 2022. Diversification is correlated with temperature in white and sulfur butterflies. *bioRxiv.*

Condamine, F. L., et al. 2023. A comprehensive phylogeny ... of *Papilio*. *Molecular Phylogenetics and Evolution.*

Lukhtanov, V. A., et al. 2005. Reinforcement of pre-zygotic isolation and karyotype evolution in *Agrodiaetus* ... *Nature.*

Pierce, N. E., and Dankowicz, E. 2022. The natural history of caterpillar-ant associations. In *Caterpillars in the middle.* Springer Int. Pub.

Tyler, H. A., et al. 1994. *Swallowtail butterflies of the Americas: ...* Scientific Pub.

Wahlberg, N., et al. 2014. Revised systematics and higher classification of pierid butterflies ... *Zoologica Scripta.*

Warren, A. D., et al. 2013. Illustrated lists of American butterflies. http://www.butterfliesofamerica.com

Chapter 6 Beran, F., and Petschenka, G. 2022. Sequestration of plant defense compounds by insects ... *Annual review of entomology.*

Godfray, H. C. J. 1994. *Parasitoids. Behavioral and Evolutionary Ecology.* Princeton Un. Pr.

James, D. G. (editor). 2018. *The book of caterpillars ...* University of Chicago Press.

Chapter 7 Grimaldi D. A. 2023. *The Complete Insect: Anatomy, Physiology, Evolution, and Ecology.* Princeton Un. Pr.

Pass, G. 2018. Beyond aerodynamics: The critical roles of the circulatory and tracheal systems in maintaining insect wing functionality. *Arthropod Str. & Dev.*

Ravenscraft, A., et al. 2019. Structure and function of the bacterial and fungal gut microbiota of Neotropical butterflies. *Ecological Monographs.*

Scoble, M. J. 1992. *The Lepidoptera ...* Oxford Un. Pr.

Snodgrass, R. E. 1961. *The caterpillar and the butterfly.* Smithsonian Miscellaneous Collns.

Srivastava, K. P. 1976. On the respiratory system of the lemon-butterfly ... *Australian Journal of Entomology.*

Tsai, C. C., et al. 2020. Physical and behavioral adaptations to prevent overheating of the living wings of butterflies. *Nature Communications.*

Chapter 8 Briscoe, A. D. 2008. Reconstructing the ancestral butterfly eye: focus on the opsins. *J of Experimental Biol.*

Chakraborty, M., et al. 2023. Sex-linked gene traffic underlies the acquisition of sexually dimorphic UV color vision ... *PNAS.*

Lane, K. A., et al. 2008. Hearing in a diurnal, mute butterfly, *Morpho peleides.* *J. of Comparative Neurology.*

McCulloch, K. J., et al. 2022. Insect opsins and evo-devo ... *Philosophical*

Transactions of the Royal Society B.

Reppert, S. M., and de Roode, J. C., 2018. Demystifying monarch butterfly migration. *Current Biol.*

Sourakov, A. A., et al. 2012. Foraging behavior of the Blue Morpho ... *Psyche.*

Sun, P., et al. 2018. In that vein: inflated wing veins contribute to butterfly hearing. *Biol. Letters.*

Chapter 9 Beldade, P., and Monteiro, A. 2021. Eco-evo-devo advances with butterfly eyespots. *Current opinion in genetics & devel.*

Ficarrotta, V., et al. 2022. A genetic switch for male UV iridescence in an incipient species pair of sulphur butterflies. *PNAS.*

Hanly, J. J., et al. 2023. Genetics of yellow-orange color variation in a pair of sympatric sulfur butterflies. *Cell Reports, 42(8).*

Jiggins, C. D. 2017. *The ecology and evolution of Heliconius butterflies.* Oxford Un. Pr.

Martin, A., and Reed, R.D., 2014. Wnt signaling underlies evolution and development of the butterfly wing pattern symmetry systems. *Devel. Biol.*

Mazo-Vargas, A., et al. 2022. Deep cis-regulatory homology of the butterfly wing pattern ground plan. *Science, 378(6617), pp.304–308.*

Pomerantz, A. F., et al. 2021. Developmental, cellular and biochemical basis of transparency in clearwing butterflies. *J of Experimental Biol.*

Sourakov, A., 2020. Emperors, admirals and giants ... *F1000Research.*

Sourakov, A. A., and Al-Obeidi, A. 2019. Biomimetic non-uniform nanostructures reduce broadband reflectivity in transparent substrates. *MRS Communications.*

Thayer, R. C., et al. 2020. Structural color in *Junonia* butterflies evolves by tuning scale lamina thickness. *Elife.*

Van Belleghem, S. M., et al. 2023. High level of novelty under the hood of convergent evolution. *Science, 379(6636), pp.1043–1049.*

Zhang, L., et al. 2017. Single master regulatory gene coordinates the evolution and development of butterfly color and iridescence. *PNAS.*

Chapter 10 Nicholls, S. 2023. *Alien Worlds: How Insects Conquered the Earth* ... Princeton Un. Pr.

Larsen, T. B. 2008. Forest butterflies in West Africa have resisted extinction ... so far ... *Biodiversity and Conservation.*

Larsen, T. B., et al. 1979. The butterfly fauna of a secondary bush locality in Nigeria. *J of Res. on the Lepidoptera.*

Reboud, E. L., et al. 2023. Genomics, Population Divergence, and Historical Demography of ... the Queen Alexandra's Birdwing. *Genome Biol. and Evol.*

Wagner, D. L., et al. 2021. Insect decline in the Anthropocene: Death by a thousand cuts. *PNAS.*

Chapter 11 Bass, M. A. 2019. *Insect Artifice: Nature and Art in the Dutch Revolt.* Princeton Un. Pr.

Daniels, J. 2022. *Your Florida Guide to Butterfly Gardening: A Guide for the Deep South.* Un. Pr. of Fla.

Impelluso, L. 2004. *Nature and Its Symbols* ... J. Paul Getty Museum.

Van Huis, A., 2019. Cultural significance of Lepidoptera in sub-Saharan Africa. *Journal of Ethnobiology and Ethnomedicine.*

Chapter 12 Carvalho, A. P. S., et al. 2017. A review of the occurrence and diversity of the sphragis in butterflies. *ZooKeys.*

Estrada, C., et al. 2010. Sex-specific chemical cues from immatures facilitate the evolution of mate guarding in *Heliconius* butterflies. *Proceedings of the Royal Society B.*

Finkbeiner, S. D. 2014. Communal roosting in *Heliconius* butterflies ... *J. of the Lepidopterists' Soc.*

Jiggins, F. M., et al. 1998. Sex ratio distortion in *Acraea encedon* is caused by a male killing bacterium. *Heredity.*

Sourakov, A. 2008. Pupal mating in *Zebra longwing* ... *News of the Lepidopterist's Soc.*

Sourakov, A. 1997. Social oviposition behavior ... of *Aglais cashmirensis. Holarctic Lepidoptera.*

Sourakov, A. 2013. Two heads are better than one: false head allows *Calycopis cecrops* to escape predation by a jumping spider ... *J. of Nat. Hist.*

Talavera, G., et al. 2023. The Afrotropical breeding grounds of the Palearctic-African migratory painted lady butterflies ... *PNAS.*

Weiss, M.R. 2003. Good housekeeping: why do shelter-dwelling caterpillars fling their frass? *Ecology Letters.*

INDEX